CHEMICAL ENGINEERING METHODS AND TECHNOLOGY

APPLICATION OF ELECTROSPUN FIBERS IN TISSUE ENGINEERING

CHEMICAL ENGINEERING METHODS AND TECHNOLOGY

Additional books in this series can be found on Nova's website under the Series tab.

Additional e-books in this series can be found on Nova's website under the e-book tab.

CHEMICAL ENGINEERING METHODS AND TECHNOLOGY

APPLICATION OF ELECTROSPUN FIBERS IN TISSUE ENGINEERING

XIUMEI MO
DAWEI LI
BENJAMIN MINDEN-BIRKENMAIER
AND
GARY L. BOWLIN

nova publishers
New York

Library of Congress Cataloging-in-Publication Data

Names: Mo, Xiumei, editor.
Title: Application of electrospun fibers in tissue engineering / editors, Xiumei Mo (College of Chemistry, Chemical Engineering and Biotechnology, Donghua University, Shanghai, China, and others) [and three others].
Description: Hauppauge, New York : Nova Science Publishers, Inc., 2016. | Series: Chemical engineering methods and technology | Includes bibliographical references and index.
Identifiers: LCCN 2015040892 (print) | LCCN 2016000013 (ebook) | ISBN 9781634636292 (hardcover) | ISBN 9781634840866 ()
Subjects: LCSH: Tissue engineering. | Nanofibers. | Fibers. | Electrospinning. | Biomedical materials.
Classification: LCC R857.T55 A67 2016 (print) | LCC R857.T55 (ebook) | DDC 571.5/38--dc23
LC record available at http://lccn.loc.gov/2015040892

Published by Nova Science Publishers, Inc. † New York

CONTENTS

PREFACE

The process of electrospinning has experienced decades of development for a variety of applications including air filtration and tissue engineering. However, it has not been until the past decade that this process has been a focus in the field of tissue engineering. The unique properties of electrospun fibers have drawn remarkable attention within the field of tissue engineering due to their capacity to mimic the native tissue structural element, the extracellular matrix, and its ability to promote cell proliferation and enhanced tissue regeneration. Herein, we have summarized some different applications of electrospun fibers in tissue engineering, while demonstrating the cellular interactions between various cells and electrospun fiber scaffolds. The key aspect is a focus on the potential use of electrospinning in tissue regeneration. The hope is that this book will be beneficial to students and researchers in developing a general knowledge of the tissue engineering and regeneration application of electrospun fibers. We hope it will also inspire their curiosity and creativity in fabricating advanced tissue engineering scaffolds to improve quality of life across the globe.

This book was supported by State Key Lab for Modification of Chemical Fibers and Polymer Materials.

Chapter 1

ELECTROSPINNING: AN INTRODUCTION

ABSTRACT

Electrospinning, a well-understood technique for the fabrication of micro- to nano-meter diameter fibers, has been extensively investigated in many fields over the past two decades. It is known for its ability to make materials with a high surface-area-to-volume ratio, high porosity, ease of surface functionalization, and simplicity in handling. More specifically, the use of electrospun fibers has drawn significant attention in the fields of high-performance filter media, composites, catalysts, energy storage, drug delivery system, and tissue engineering. The area of greatest interest over the last decade has been the application of electrospun fibers in the field of tissue engineering. These fibers are used to create extracellular matrix analogues onto which cellular components are seeded in an attempt to regenerate three-dimensional tissues and organs. For this reason, this publication will focus on the process of electrospinning and its widespread use in tissue engineering.

Keywords: Electrospinning, tissue engineering, nanofibers, scaffolds

1. INTRODUCTION

The electrospinning process was discovered well over a century ago, with the first patent issued on this technology in 1934. However, probably due to scale-up issues, this technique never gained widespread traction and has only recently gained popularity. At its most basic, the electrospinning process utilizes an electric field to generate a polymer jet from a solution, forming fibers that can be collected to form non-woven structures (Reneker et al.). One of the simplest and most basic of the electrospinning devices used is illustrated in Figure 1. The three key components of such a basic system are a high voltage power supply (preferably current limited for safety), a syringe pump for metering the polymer solution delivery rate to the nozzle (typically a blunt-end needle), and a grounded collector system (the simplest form being a section of aluminum foil). The polymer, which is typically maintained in a solution but can be in a melt, is delivered by the syringe pump through the nozzle orifice at a controlled flow rate. A charge imbalance is then induced in the polymer solution by using a power supply to apply a high voltage to the metallic nozzle. This causes a repulsive interaction between like charges in the solution while at the same time setting up a static electric field between the polymer solution surface and the grounded collection system. As the electric field strength is increased, the solution surface tension is overcome and the solution begins to be drawn out of the nozzle until a Taylor cone is formed. This occurs when the electrical forces in solution are balanced by surface tension. A charged polymer jet will be ejected from the tip of the Taylor cone while the electric field forces are higher than surface tension. The formed jet is drawn and accelerated towards the grounded collector, during which the diameter of the jet is reduced and the solvent evaporates to form polymeric fibers collected on the grounded target. Images of electrospun structures are shown in Figure 2.

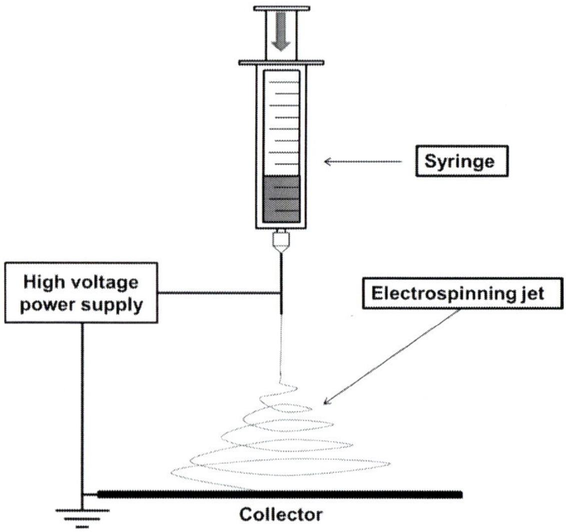

Figure 1. Schematic of electrospinning.

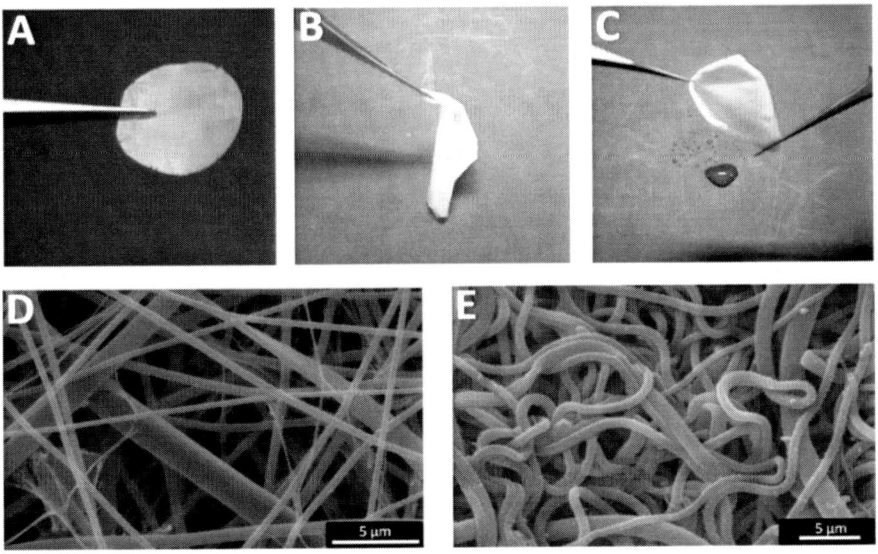

Figure 2. Electrospun elastin structures. (A) uncrosslinked structure, (B) crosslinked with hexamethylene diisocyanate, (C) wetted with phosphate buffered saline, (D) scanning electron micrograph of uncrosslinked fibers, (E) crosslinked structure (© 2011 Nivison-Smith L, Weiss A. Published in Nivison-Smith et al. under CC BY-NC-SA 3.0 license. Available from: http://dx.doi.org/10.5772/23660. Reprinted with permission).

In terms of fiber processing, there are several controllable parameters which affect the formation of fibers during electrospinning. The solution parameters, including conductivity, surface tension, and viscosity, determine the amount of stretching of the solution, playing a role in its ability to be electrospun. The most widely controlled parameter is viscosity, which is directly proportional to polymer concentration and molecular weight. Increasing viscosity increases the fiber diameter, and vice versa. Additionally, higher surface tension and viscosity generally makes it more difficult to generate a Taylor cone and the stretching of the polymer jets. The evaporation rate of the solvent also influences the viscosity of the solution as the polymer jet is being stretched (Ramakrishna et al.). Another possibility at low viscosities is that as the jet travels to the collector, the charge repulsion within the solution may overcome the solution surface tension, creating beads instead of continuous fibers. Moreover, as the polymer jet is stretched by the electric field, the amount of charge carried by the jet will influence the electrical forces encountered, contributing to a whipping motion during fiber travel. It has also been reported that an increase in solution conductivity will result in smaller diameter fibers being formed (Zong et al.). However, this increase in solution conductivity may also lead to greater bending instability, which increases the deposition area of the fibers (Choi et al.).

The major process parameters of applied voltages, collecting distance, flow rate of the solution, and electrical field strength affect the collected fiber morphology. The crucial driving element in electrospinning is the application of a high voltage to the solution. As noted, the high voltage introduces polarization within the solution, generating the Taylor cone from which polymer solution jets are produced and subsequently drawn into fibers. The applied voltage magnitude affects the stability of the Taylor cone. Generally, a higher applied voltage is needed to keep the Taylor cone stable when the feed rate of solution is increased (Zong et al.). The applied voltage and the generated electric field also influence the stretching and acceleration of the polymer jet. Typically, higher voltage introduces greater stretching of the jet due to the greater electric field strength, resulting in smaller diameter fibers (Lee et al.). The applied voltage also has an influence on the crystallinity of the formed polymer fiber, with stronger electrostatic fields causing the polymer molecules to arrange in a more ordered fashion during the jet formation. With the help of auxiliary electrodes or changing the shape of the collector, the electric field can be regulated, allowing one to obtain aligned and even patterned fibrous non-woven scaffolds (Vaquette and Cooper-White; Simsek et al.). For a set voltage, an increase in the polymer solution feed rate will induce a corresponding increase in the resultant fiber diameter or bead size (Rutledge et al.). As for the collecting distance, it not only determines the electrostatic field strength, but it also has an effect on the flight time which determines the degree of solvent evaporation. The results of utilizing a low collecting distance will typically not allow for sufficient solvent evaporation which results in wet fibers collecting that fuse together where they intersect and interact (Buchko et al.). This welded morphology can also be achieved through melt-welding, a procedure in which the electrospun fibers are exposed to a vapor-phase polymerization step which increases and strengthens the connections between the fibers. A typical image of melded fibers is shown in Figure 3.

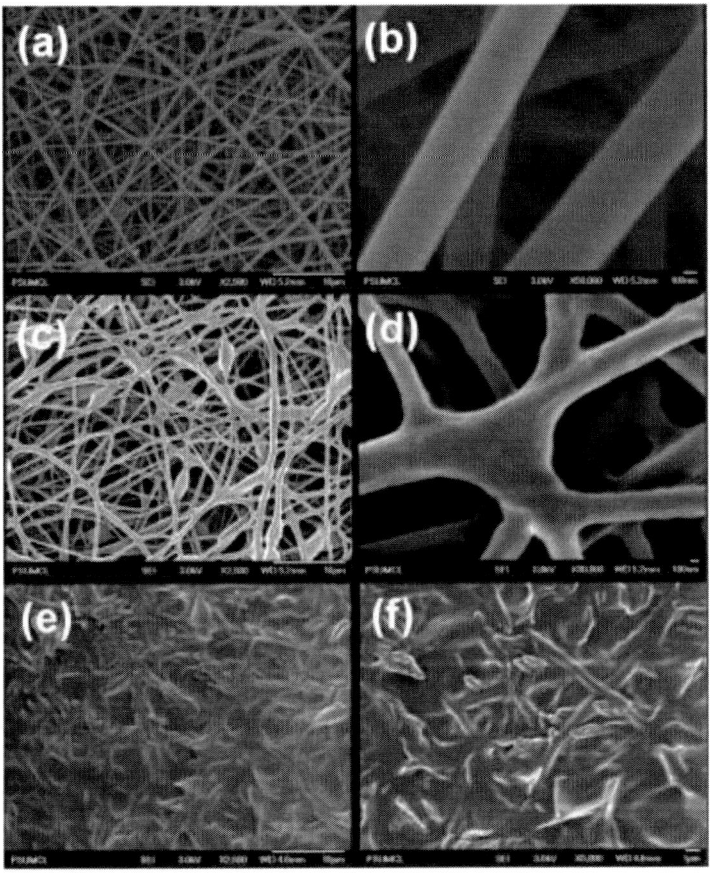

Figure 3. Melt-welded nanofibers spun from Poly-3,4-ethylenedioxythiophene. (Nair et al. Reprinted with permission from the American Chemical Society).

1.1. Electrospun Fiber

Electrospun fiber morphology, structure, and arrangement are functions of the electrospinning parameters. A key feature of this process is that it is capable of fabricating diverse types of fibers with interesting morphologies. For example, changing the ambient processing environmental humidity will result in fibers with a porous surface (Figure 4) (Bognitzki et al.). Under certain processing conditions related to solvent evaporation, ribbon-like fibers (Figure 5) can be generated (Koombhongse, Liu and Reneker). Another popular permutation is the formation of core-shell fibers (Figure 6) by emulsion electrospinning and co-axial electrospinning (Viry et al., X. Li et al.) which have tremendous possibilities in the field of drug delivery. Additionally, the use of co-axial electrospinning makes it possible to fabricate bio-component fibers (Huang et al.).

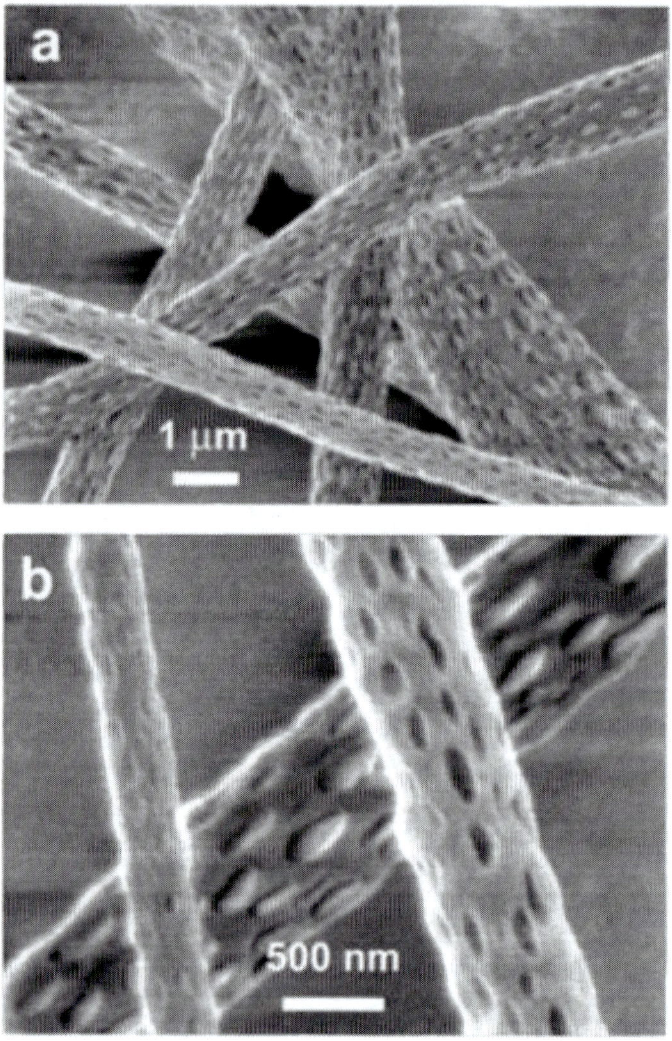

Figure 4. Scanning electron micrographs of poly(L-lactic acid) fibers nanopatterned with a porous surface. (Bognitzki et al. Reprinted with permission from John Wiley and Sons).

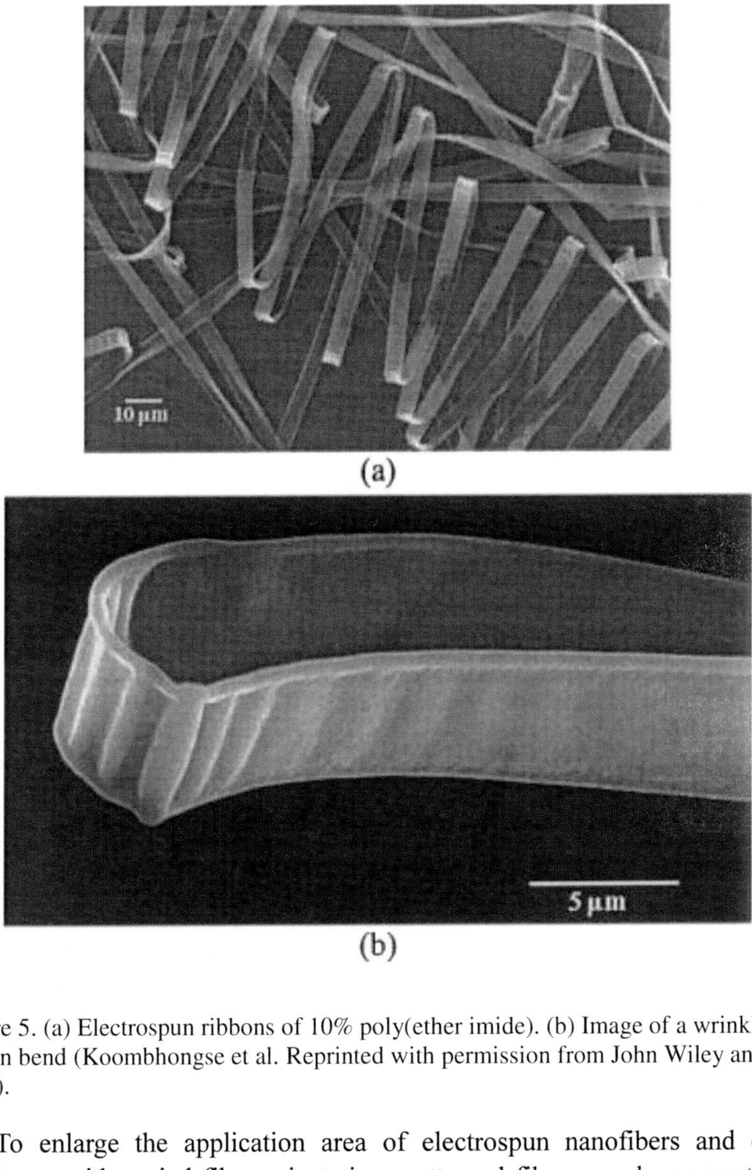

Figure 5. (a) Electrospun ribbons of 10% poly(ether imide). (b) Image of a wrinkled ribbon bend (Koombhongse et al. Reprinted with permission from John Wiley and Sons).

To enlarge the application area of electrospun nanofibers and create structures with varied fiber orientation, patterned fibers can be generated by modification of the electrospinning setups. These structures of one-, two-, and three-dimensional fiber architectures with controlled alignments can be obtained with customized processing and specified parameters. A simple variation includes a rotating cylindrical collection mandrel with high rotational speeds that guides fiber alignment into a unidirectional structure (Boland et al.). Another potential process variation is the shape of the collector used. The collector shape affects the distribution of electrostatic field significantly, which affects subsequent fiber deposition. With this knowledge, researchers have designed systems in which there are two collectors arranged in parallel to allow the collecting fibers to bridge between the two electrodes and arrange in a parallel, highly aligned fashion (Figure 7) (Katta et al., Dersch et al.).

Another classical example utilized a rotating disc with a sharp edge to create parallel aligned fibers (Figure 8) (Theron et al.). Yet another twist on the process is the use of a water bath collector or other engineered collecting systems that allow for the collecting fibers to be twisted and form yarns from the electrospun fibers (Figure 9) (Hosseini Ravandi et al., Sun; J. Wu et al., Reneker and Chun; Yan, Liu and Zhang, Afifi et al.).

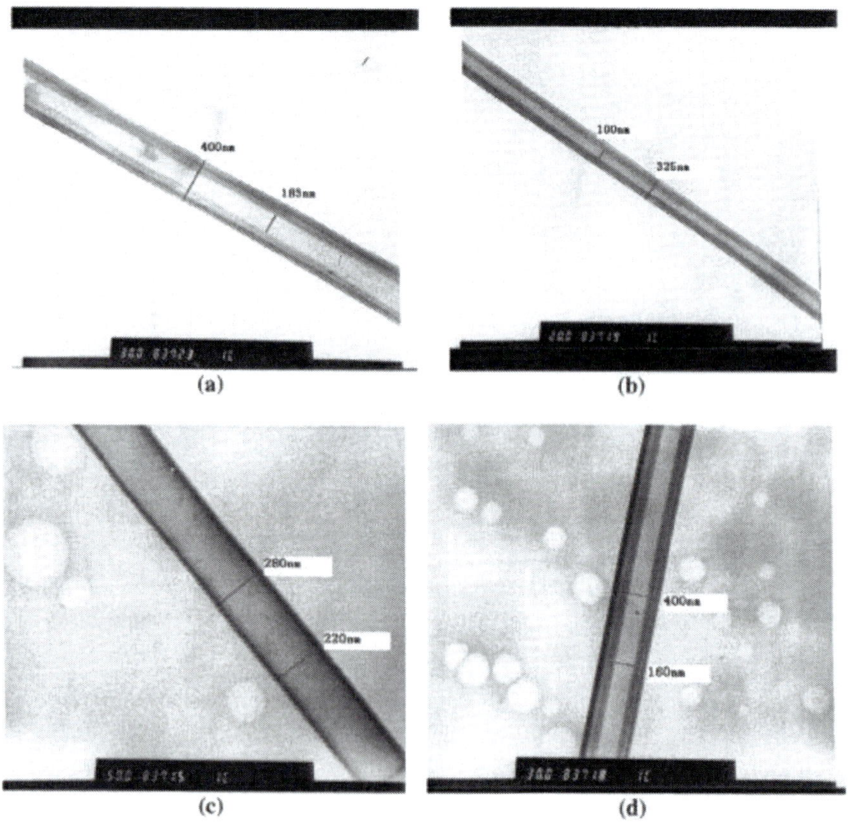

Figure 6. Transmission electron micrograph images of core-shell fibers spun from polycaprolactone (Huang et al. Reprinted with permission from John Wiley and Sons).

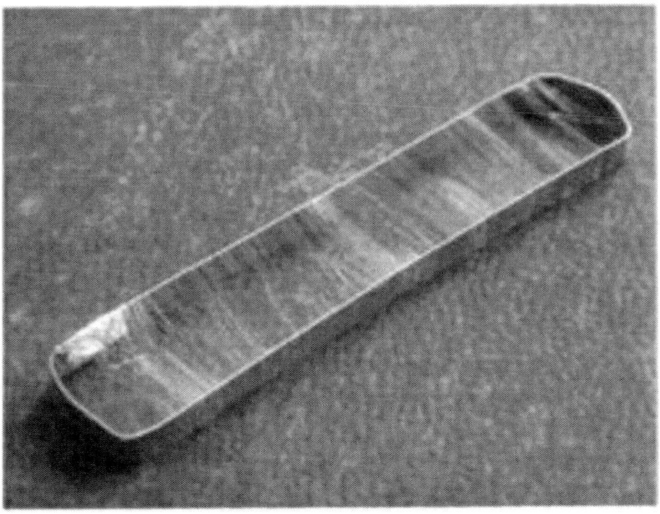

(a)

(b)

Figure 7. (a) Electrode used to create parallel fibers. (b) Parallel electrospun poly(lactic acid) fibers (Dersch et al. Reprinted with Permission from John Wiley and Sons).

Figure 8. Plexiglass disk covered by parallel copper wires for the collection of aligned electrospun fibers (Katta et al. Reprinted with permission from the American Chemical Society).

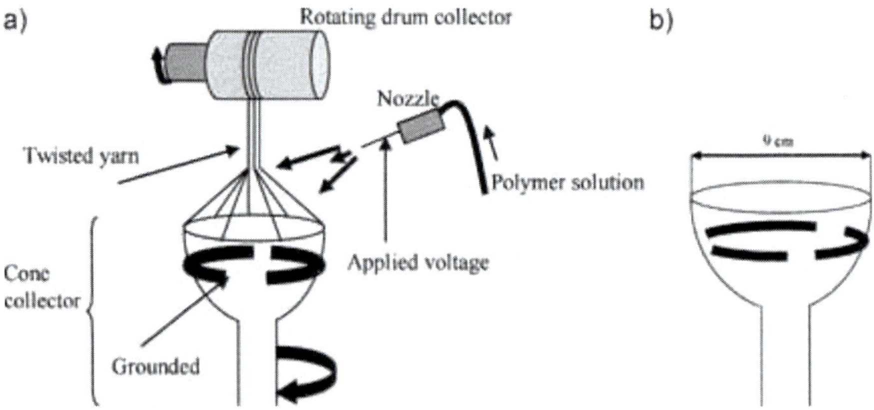

Figure 9. Schematic of method used to create electrospun yarn. (A) The electrospinning setup, and (B) the rotating funnel collector used to turn the fibers into yarn. (Afifi et al. Reprinted with permission from John Wiley and Sons)

1.2. Materials for Electrospinning

Electrospinning is the most versatile and reliable method to fabricate continuous micro- to nano-fiber structures compared to methods such as phase separation and self-assembly. Not only is the process simple, it is versatile due to the ability to process a wide range of materials such as synthetic materials, natural polymers, and ceramic into various morphologies or structures on the micro- to nano-meter scale (Goh, Shakir and Hussain). In the last decade, electrospinning has been intensely investigated because of this versatility. For tissue engineering applications, commonly used synthetic polymers consist of polycaprolactone (PCL) (Cipitria et al.), polyacrylonitrile (PAN) (Nataraj, Yang and Aminabhavi), poly-L-lactide (PLLA) (Blackwood et al.), polyglycolide (PGA) (Song et al.), polyvinyl-alcohol (PVA) (Vaquette et al., C. Wu et al.), poly(L-lactide-co-caprolactone) (P(LLA-CL)) (Xin, Hussain and Mao; Kim et al.), poly(D,L-lactide-co-glycolide) (PLGA) (Xin, Hussain and Mao; Kim et al.), and polydioxanone (PDO) (Boland et al.). Natural materials processed by electrospinning include collagen (Matthews et al. 2003, Erisken et al.), gelatin (Y. Zhang et al.), fibrinogen (Wnek et al.), silk fibroin (Zhang, Baughman and Kaplan), chitin (Noh et al.), chitosan (Geng, Kwon and Jang), and cellulose acetate (Han et al., J. Li et al.). To overcome the drawbacks of the synthetic and natural polymers as individual fiber components, many groups have resorted to blending multiple polymers in solution for electrospinning to create novel fibers with unique properties (Chiu et al.). Moreover, nanoparticles of inorganic materials such as hydroxyapatite (HA) (Rodenas-Rochina, Ribelles and Lebourg), silver (Patel et al.), platinum (Formo et al.), and iron oxide (D. Zhang et al.) can be encapsulated within the fibers during the process.

1.3. Perspective

While a great deal of emphasis has been placed on the utilization of electrospun fibers in tissue engineering scaffolds, there remains a great deal of research and development needed to create the variety of scaffolds necessary for tissue specific regeneration. This is because our tissues and organs display a specificity and great diversity in the native (very much organ and tissue specific) extracellular matrices we are trying to mimic. This diversity highlights the fact that one fiber size or composition will not work for every application, and each application will be highly specific. Thus, the versatility of the electrospinning process and its ability to create micro- to nano-scale fibers from a wide variety of both synthetic and natural polymers as well as other additives will keep it at the forefront of scaffold development for many years to come.

REFERENCES

Afifi et al. "Electrospinning of Continuous Aligning Yarns with a 'Funnel' Target." *Macromolecular Materials and Engineering* 295 (2010): 660-665.

Blackwood, K.A., et al. "Development of Biodegradable Electrospun Scaffolds for Dermal Replacement." *Biomaterials* 29 (2008): 3091-104.

Bognitzki, M., et al. "Nanostructured Fibers Via Electrospinning." *Advanced Materials* 13 (2001): 70-72.

Boland, E.D., et al. "Tailoring Tissue Engineering Scaffolds Using Electrostatic Processing Techniques: A Study of Poly (Glycolic Acid) Electrospinning." *Journal of Macromolecular Science, Part A* 38 (2001): 1231-43.

Boland, E.D., et al. "Electrospinning Polydioxanone for Biomedical Applications." *Acta Biomaterialia* 1 (2005): 115-123.

Buchko, C.J., et al. "Processing and Microstructural Characterization of Porous Biocompatible Protein Polymer Thin Films." *Polymer* 40 (1999): 7397-407.

Chiu, J.B., et al. "Functionalization of Poly(L-lactide) Nanofibrous Scaffolds with Bioactive Collagen Molecules." *Journal of Biomedical Materials Research Part A* 83 (2007): 1117-27.

Choi, J.S., et al. "Effect of Organosoluble Salts on the Nanofibrous Structure of Electrospun Poly (3-Hydroxybutyrate-co-3-Hydroxyvalerate)." *International Journal of Biological Macromolecules* 34 (2004): 249-56.

Cipitria, A., et al. "Design, Fabrication and Characterization of PCL Electrospun Scaffolds—a Review." *Journal of Materials Chemistry* 21 (2011): 9419-53.

Dersch, R., et al. "Electrospun Nanofibers: Internal Structure and Intrinsic Orientation." *Journal of Polymer Science Part A: Polymer Chemistry* 41 (2003): 545-53.

Erisken, C., et al. "Scaffold Fiber Diameter Regulates Human Tendon Fibroblast Growth and Differentiation." *Tissue Engineering Part A* 19 (2013): 519-28.

Formo, E., et al. "Functionalization of Electrospun TiO2 Nanofibers with Pt Nanoparticles and Nanowires for Catalytic Applications." *Nano Letters* 8 (2008): 668-72.

Geng, X., Oh-Hyeong, K., and J. Jang. "Electrospinning of Chitosan Dissolved in Concentrated Acetic Acid Solution." *Biomaterials* 26 (2005): 5427-32.

Goh, Y.F., Shakir, I., and R. Hussain. "Electrospun Fibers for Tissue Engineering, Drug Delivery, and Wound Dressing." *Journal of Materials Science* 48 (2013): 3027-54.

Han, S.O., et al. "Electrospinning of Cellulose Acetate Nanofibers Using a Mixed Solvent of Acetic Acid/Water: Effects of Solvent Composition on the Fiber Diameter." *Materials Letters* 62 (2008): 759-62.

Hosseini Ravandi, S. A., et al. "Mechanical Properties and Morphology of Hot Drawn Polyacrylonitrile Nanofibrous Yarn." *Journal of Applied Polymer Science* (2011): 5002-5009.

Huang et al. "Encapsulating Drugs in Biodegradable Ultrafine Fibers Through Co-Axial Electrospinning." *Journal of Biomedical Materials Research* 77A (2006): 169-179

Katta, P., et al. "Continuous Electrospinning of Aligned Polymer Nanofibers onto a Wire Drum Collector." *Nano Letters* 4 (2004): 2215-18.

Kim, S.J., et al. "Fabrication and Characterization of 3-Dimensional PLGA Nanofiber/Microfiber Composite Scaffolds." *Polymer* 51 (2010): 1320-27.

Koombhongse, S., Liu, W. and D.H. Reneker. "Flat Polymer Ribbons and Other Shapes by Electrospinning." *Journal of Polymer Science Part B: Polymer Physics* 39 (2001): 2598-606.

Lee, J.S., et al. "Role of Molecular Weight of Atactic Poly (Vinyl Alcohol)(PVA) in the Structure and Properties of PVA Nanofabric Prepared by Electrospinning." *Journal of Applied Polymer Science* 93 (2004): 1638-46.

Li, J., et al. "Electrospinning of Hyaluronic Acid (HA) and HA/Gelatin Blends." *Macromolecular Rapid Communications* 27 (2006): 114-20.

Li, X., et al. "Encapsulation of Proteins in Poly(L-Lactide-Co-Caprolactone) Fibers by Emulsion Electrospinning." *Colloids and Surfaces B: Biointerfaces* 75 (2010): 418-24.

Matthews, J.A. et al. "Electrospinning of Collagen Type II: A Feasibility Study." *Journal of Bioactive and Compatible Polymers* 18 (2003): 125-34.

Matthews, J.A. et al. "Electrospinning of Collagen Nanofibers." *Biomacromolecules* 3 (2002): 232-238.

Nair, S., Hsiao, E., and S.H. Kim. "Melt-Welding and Improved Electrical Conductivity of Nonwoven Porous Nanofiber Mats of POly(3,4-

ethylenedioxythiophene) Grown on Electrospun Polystyrene Fiber Template." *Chemistry of Materials* 21 (2009): 115-121.

Nataraj, S.K., Yang, K.S., and T.M. Aminabhavi. "Polyacrylonitrile-Based Nanofibers—a State-of-the-Art Review." *Progress in Polymer Science* 37 (2012): 487-513.

Nivison-Smith, et al. *Elastin Based Constructs, Regenerative Medicine and Tissue Engineering - Cells and Biomaterials.* InTech, 2011.

Noh, H.K., et al. "Electrospinning of Chitin Nanofibers: Degradation Behavior and Cellular Response to Normal Human Keratinocytes and Fibroblasts." *Biomaterials* 27 (2006): 3934-44.

Patel, A.C., et al. "Electrospinning of Porous Silica Nanofibers Containing Silver Nanoparticles for Catalytic Applications." *Chemistry of Materials* 19 (2007): 1231-38.

Ramakrishna, S., et al. *An Introduction to Electrospinning and Nanofibers.* Vol. 90: World Scientific, 2005.

Reneker, D. H., et al. "Electrospinning of Nanofibers from Polymer Solutions and Melts." *Advances in Applied Mechanics*. Eds. Aref, H. and E. van der Giessen. Vol. 41: Elsevier, 2007. 345-346.

Reneker, D.H, and I. Chun. "Nanometre Diameter Fibres of Polymer, Produced by Electrospinning." *Nanotechnology* 7 (1996): 216-223.

Rodenas-Rochina, J., Ribelles, J.L., and M. Lebourg. "Comparative Study of PCL-HAP and PCL-Bioglass Composite Scaffolds for Bone Tissue Engineering." *Journal of Material Science: Materials in Medicine* 24 (2013): 1293-308.

Rutledge, G.C., et al. "Electrostatic Spinning and Properties of Ultrafine Fibers." *Annual Report (M98-D01)* (2001): 1-10.

Simsek, M., et al. "Chitosan and Polycaprolactone Membranes Patterned Via Electrospinning: Effect of Underlying Chemistry and Pattern Characteristics on Epithelial/Fibroblastic Cell Behavior." *Journal of Biomedical Materials Research Part A* 100 (2012): 3332-43.

Song, W., et al. "Electrospun Polyvinyl Alcohol-Collagen-Hydroxyapatite Nanofibers: A Biomimetic Extracellular Matrix for Osteoblastic Cells." *Nanotechnology* 23 (2012): 115101.

Sun, F., et al. "Fabrication of Poly(Vinylidenefluoride-Co-Hexafluoropropylene) Nanofiber Yarns by Conjugate Electrospinning." *Journal of the Textile Institute* 102 (2011): 633-38.

Theron, A., Zussman, E., and A.L. Yarin. "Electrostatic Field-Assisted Alignment of Electrospun Nanofibres." *Nanotechnology* 12 (2001): 384.

Vaquette, C., et al. "Aligned Poly(L-Lactic-Co-E-Caprolactone) Electrospun Microfibers and Knitted Structure: A Novel Composite Scaffold for Ligament Tissue Engineering." *Journal of Biomedical Materials Research Part A* 94 (2010): 1270-82.

Vaquette, C. and J.J. Cooper-White. "Increasing Electrospun Scaffold Pore Size with Tailored Collectors for Improved Cell Penetration." *Acta Biomaterialia* 7 (2011): 2544-57.

Viry, L., et al. "Emulsion-Coaxial Electrospinning: Designing Novel Architectures for Sustained Release of Highly Soluble Low Molecular Weight Drugs." *Journal of Materials Chemistry* 22 (2012): 11347.

Wnek, G.E., et al. "Electrospinning of Nanofiber Fibrinogen Structures." *Nano Letters*, 3 (2003): 213-16.

Wu, C., et al. "A Novel Heparin Loaded Poly(L-Lactide-Co-Caprolactone) Covered Stent for Aneurysm Therapy." *Materials Letters* 116 (2014): 39-42.

Wu, J., et al. "Cell Infiltration and Vascularization in Porous Nanoyarn Scaffolds Prepared by Dynamic Liquid Electrospinning." *Journal of Biomedical Nanotechnology* 10 (2014): 603-14.

Xin, X., Hussain, M. and J.J. Mao. "Continuing Differentiation of Human Mesenchymal Stem Cells and Induced Chondrogenic and Osteogenic Lineages in Electrospun PLGA Nanofiber Scaffold." *Biomaterials* 28 (2007): 316-25.

Yan, H., Liu, L., and Z. Zhang. "Continually Fabricating Staple Yarns with Aligned Electrospun Polyacrylonitrile Nanofibers." *Materials Letters* 65 (2011): 2419-21.

Zhang, D., et al. "Electrospun Polyacrylonitrile Nanocomposite Fibers Reinforced with Fe3O4 Nanoparticles: Fabrication and Property Analysis." *Polymer* 50 (2009): 4189-98.

Zhang, X., Baughman, C.B., and D.L. Kaplan. "*In vitro* Evaluation of Electrospun Silk Fibroin Scaffolds for Vascular Cell Growth." *Biomaterials* 29 (2008): 2217-27.

Zhang, Y., et al. "Electrospinning of Gelatin Fibers and Gelatin/Pcl Composite Fibrous Scaffolds." *Journal of Biomedical Materials Research Part B Applied Biomaterials* 72 (2005): 156-65.

Zong, X., et al. "Structure and Process Relationship of Electrospun Bioabsorbable Nanofiber Membranes." *Polymer* 43 (2002): 4403-12.

Chapter 2

INTERACTIONS BETWEEN CELLS AND ELECTROSPUN SCAFFOLDS

ABSTRACT

Electrospinning is a simple, effective method of producing micro- to nano- fibers of various compositions and morphologies. Electrospinning has become vastly popular in the tissue engineering community as a method to produce scaffolds which closely mimic the architecture and composition of native tissue extracellular matrix. There are a variety of studies indicating that the micro- and nano-topology of the electrospun scaffolding significantly affects the behavior of cells cultured on/in the newly regenerated tissues. For a given tissue or organ, scaffolding of specific architecture and composition will be needed. In this chapter, we have focused on the critical interaction between cells and electrospun fibrous structures of different morphologies, including various types of fiber alignment, fiber diameter, pore size, and fiber surface topology. This chapter summarizes the performance of these scaffold as it relates to cellular interactions and their ability to regenerate various tissues and organs.

Keywords: electrospinning, cell interaction, cell adhesion, cell migration, cell differentiation

INTRODUCTION

Tissue engineering is an interdisciplinary field that employs the basic principles and innovations from engineering, material science, and the life sciences to develop functional tissue or organ replacements that will maintain or restore functionality after trauma or disease. The overall paradigm of this field is the combining of a scaffold, cells, and stimuli (mechanical forces and/or biologically active molecules) for the development of these functional tissues or organs. This is often referred to as the tissue engineering triad. The aspect that distinguishes this approach is the use of an often temporary, biodegradable scaffolding as an extracellular matrix analogue to support three-dimensional tissue regeneration employing the other triad elements.

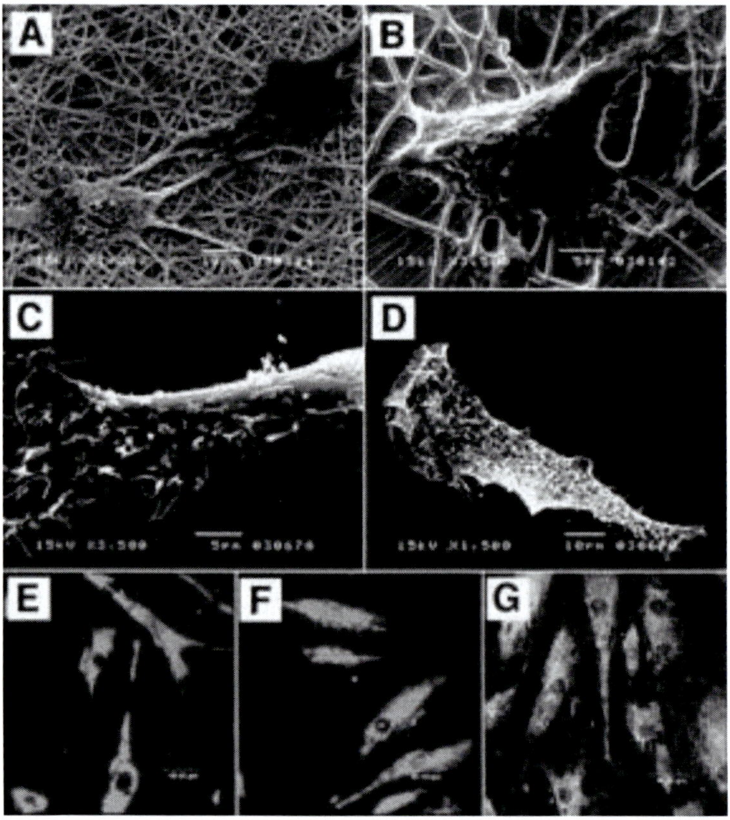

Figure 10. Scanning electron microscopy images of cells cultured on electrospun nanofibers (A,B) and tissue culture plastic (C,D). Confocal images of cells cultured on nanofibers with diameters of 50-200 nm (E) 200-500nm (F) and 500-1000 nm (G). (Shih et al. Reprinted with permission from John Wiley and Sons).

The extracellular matrix is a complex arrangement of proteins and polysaccharides, including collagen, hyaluronic acid, proteoglycans, glycosaminoglycans, and elastin. The structural elemental fibers composing the extracellular matrix generally range from 50 to 300 nm in diameter (Piez et al.). Tissue and organ specific cells are embedded in this nanofibrous gel network, forming body tissues (Ingber et al.). The extracellular matrix components are constantly synthesized, secreted, oriented, and modified by the cellular components that they support. Not that long ago, the extracellular matrix function was believed to be solely a structural one, maintaining shape and form of the tissues and organs. However, studies have shown that the extracellular matrix plays a vital role in controlling cell interactions within its three-dimensional microenvironment. One way this is accomplished is through fiber interaction with receptors on the cell membrane surface which directly promote cell adhesion, migration, growth, differentiation, and apoptosis. The extracellular matrix also plays a role in cytokine activity and intracellular signaling, both of which serve numerous functions in cell regulation and activation (Martins-Green, Palsson and Bhatia). This communication is dynamic and reciprocal between the cells and the extracellular matrix.

The aim of tissue engineering scaffolds is to mimic the extracellular matrix, providing a temporary structural element to promote and guide cellular interactions (i.e., migration) to facilitate the formation of new tissue or organs (Figure 10). The structural degradation rate should also be tailored to avoid interfering with regeneration or inducing adverse effects upon implantation. In general, there are several design and material criteria that must be met to fabricate a successful scaffold. First and foremost, the scaffold material needs to be biocompatible, meaning that it does not initiate any adverse tissue development or immune reactions. It is widely believed that the materials used in fabrication should be biodegradable or bioabsorbable at a defined rate, which allows for gradual incorporation into the surrounding neo-tissue without any fibrous encapsulation or residual evidence of their presence. Ideally, this scaffold would mimic the extracellular matrix architecture on the macro-, micro-, and even nano-scale, as each will likely have a role in influencing cellular responses. The electrospinning process allows for the creation of extracellular matrix-replicating scaffolds composed of micro- to nano-scale fibers. In addition, the morphological similarities between the nanofibrous structures and the native extracellular matrix are believed to improve cellular response, which is the focus of this chapter.

Research has demonstrated that the chance of scaffold failure will decrease as fiber diameter decreases from micro- to nano-fiber (Kim and Mooney). In support of these findings, an additional study demonstrated that electrospun scaffolds composed of micro-scale and smaller diameter fibers resulted in little-to-no fibrous encapsulation when compared to larger diameter fibers (Sanders). More specifically, researchers have also shown that the nanofibrous structure improves the use of these scaffolds when applied to bone, cartilage, cardiovascular, nervous and bladder regeneration with less scar formation (Patch and Body Smalley, Muzzarelli et al.).

In terms of direct cellular interactions with electrospun scaffolds, researchers have shown that materials of nano-scale could have a significant effect on cellular behavior. Compared with micro-scale fibrous scaffolds, cells had increased adhesion and proliferation, and secreted more collagen and elastin on nano-scale scaffolds (Pattison et al.). On the other hand, the high surface area of the nano-scale scaffolds could result in the adsorption of more proteins, resulting in more adhesion sites for the cell membrane receptors. These adsorbed proteins could translate to an increase in adhesion sites by transforming cell conformation and facilitating adhesion and proliferation. It is also known that cell viability, proliferation, and morphology of anchorage-dependent cells are directly related to the micro- and nano-scale physical features of the scaffold surface, which may be modulated by the control of the electrospinning process (Li et al.). Thus, scaffolds made of nanofibers appear to mimic the physical structure of the extracellular matrix, which may lead to the promotion of tissue regeneration. The physical scaffolding aspects to be considered in this chapter are fiber alignment, fiber diameter, surface topology of fibers, and the pore size.

2.1. Fiber Alignment

In general, electrospun scaffolds spun onto a traditional static collector exhibit randomly orientated fibers in a non-woven structure. However, by utilizing systems such as a high speed rotating cylinder, rotating disc, or two parallel grounded collectors, electrospun fibers with a high degree of alignment can be obtained. The degree of alignment can be easily quantified, allowing for reproducibility of experimental results using aligned electrospun fibers (Ayres et al.). Scaffold fiber alignment is important in many tissue engineering applications, such as neural regeneration, with a vast array of studies indicating that the fiber alignment of electrospun scaffolds affects the overall behavior of interacting cells (Chew et al., Bashur et al., Sangsanoh et al., Simsek et al., Lee et al., Yin et al., Kai et al.). In addition, electrospun fibers with an aligned orientation always have greater strength than random fibers in the direction of their alignment, and lower strength orthogonal to fiber alignment due to the low number of entanglements. In terms of cellular interactions, Chew et al. reported that human Schwann cells cultured on randomly oriented polycaprolactone fibers had a randomly oriented morphology extending across multiple fibers, and trended towards elongation along the local fiber axes. In contrast, when cultured on aligned fiber structures, all cells were aligned and elongated along the major fiber axis. These aligned electrospun fibers also enhanced the maturation of Schwann cells to a greater degree than the randomly oriented fibers. In another study, it was demonstrated that a polycaprolactone/collagen electrospun fiber structure with a unidirectional fiber orientation drove the alignment of human skeletal muscle cells along the major axis, and enhanced the formation of myotubes (Choi et al.). Overall, these aligned polycaprolactone/collagen electrospun fibers significantly facilitated the organization of skeletal muscle cells into the native tissue-like structure that is necessary to improve muscle function and maturation. Neonatal mouse cerebellum C17.2 stem cells have also been shown to demonstrate a great dependence on fiber alignment. These cells were induced to grow along the unidirectional axis of aligned fibers with the formation of focal contacts (He et al.). In yet another example utilizing electrospun aligned fiber scaffolds, longer neurite lengths were obtained versus those of cells on random fibers due to directional contact guidance performed by the aligned fibers (Figure 11). As noted, the mechanical properties of the aligned fiber structures were related to the orientation of fibers, with the aligned fibers being anisotropic and having increased stiffness in the direction of fiber alignment. One group has demonstrated that the spreading of mammalian cells is reduced on less rigid materials. Thus, cells may preferentially align parallel to the more rigid axis (Discher, Janmey and Wang).

Contact guidance is considered as the main reason for the elongation and alignment of cells on aligned electrospun fiber scaffolds. The effect of this contact guidance induced by the aligned fibers appears to be more dramatic than the randomly oriented fibers. When cultured on aligned fibers, the cytoskeleton and nuclei align and elongate in the direction of the major fiber axis. This degree of cellular alignment has also been demonstrated with other cell types such as human fibroblasts echoing the Schwann cells in other studies (Sell et al., Chew et al., Lee et al., Bashur, Dahlgren and Goldstein). However, highly aligned electrospun fibers may not be beneficial for the development of ligament tissue or three-dimensional tissue composed of highly oriented cells without significant processing advancements. The reason for this disconnect is that the packing density of these formed scaffolds is relatively efficient with this increase in fiber alignment, which reduces the overall structure's pore size and porosity, limiting the pore area (interconnected void space) available for cellular infiltration (Baker et al.).

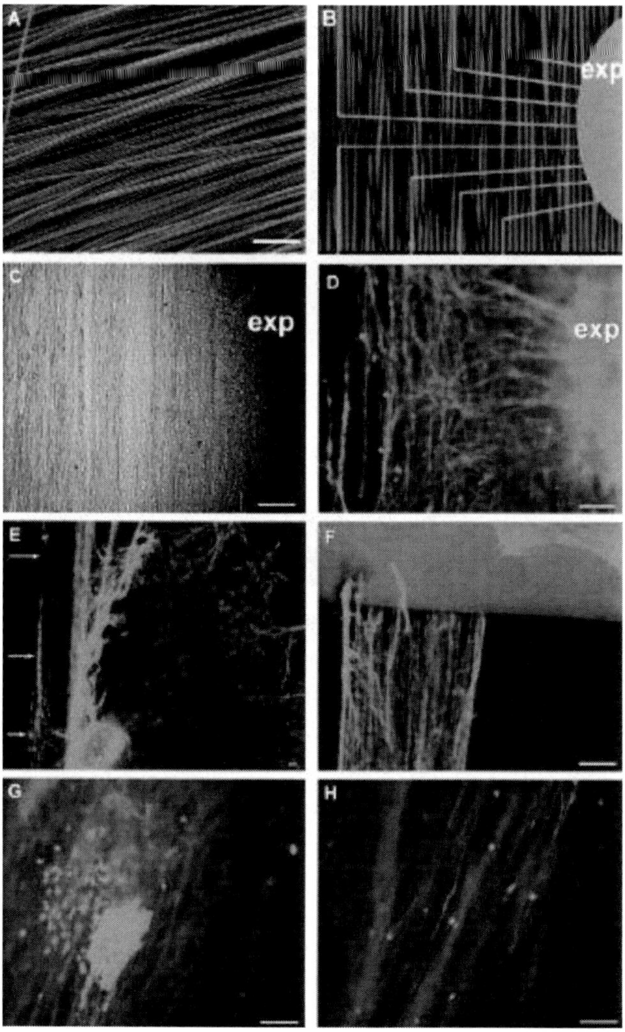

Figure 11. Scanning electron microscopy images of (A) aligned poly-L-lactate electrospun fibers, (B) ideal behavior of neurite extension from rat dorsal root ganglia (DRG) explants, (C) DRG explant cultured on aligned poly-L-lactate fibers, (D) neurites with stained neurofilaments extending from an explant to contact collagen-coated poly-L-lactate fibers and then follow their vertical alignment, (E) DRG explant on glass extends neurites radially on glass on one side while on the other side, the neurites follow the alignment of a fiber bundle, (F) neurites do not leave the fiber bundle to grow in anchoring tape or glass, (G) after 12 days, the DRG explant cell bodies degrade, but the neurites remain attached to fibers, (H) cell bodies can be found over 100 μm away from explant seen in G, showing migration along the fibers. (Corey et al. Reprinted with permission from John Wiley and Sons).

2.2. Fiber Diameter

Cell bchavior on electrospun fibers is also affected by the fiber diameter, which can be adjusted by processing parameters and solution properties such as electric field strength, collecting distance, and polymer solution concentration. It has been shown that the diameter of electrospun fibers significantly influences cell adhesion, spreading/morphology, proliferation, and the ability of cells to migrate on the substrates (Ramakrishna et al., Ming Chen et al., Sisson et al., Christopherson, Song and Mao; Erisken et al., Thompson et al.). Moreover, the influence of fiber diameter on cell behavior varies with cell type.

Sisson et al. demonstrated that electrospun scaffold fiber diameter also affects the differentiation of osteoblastic MG63 cells (Sisson et al.). Their results indicated that the MG63 cell line proliferated more over time and migrate further into the large fiber diameter scaffolds, but differentiate faster on the small fiber diameter scaffolds on which they more rapidly activated focal adhesion kinase (Figure 12). Neural stem cells cultured on fibers with larger diameters supported lower migratory activity and developed large aggregates (Christopherson, Song and Mao). More specifically, neural stem cells showed reduced migration, spreading, and proliferation as the electrospun fiber diameter increased. While in the differentiation medium, the neural stem cells spread and assumed glial cell shape, and preferentially differentiate into oligodendrocytes. However, on 749 and 1452 nm fibers they elongated and preferentially differentiated into a neuronal lineage. For Schwann cells, small diameter fibers (293 ± 65 nm) did not promote extensive neurite extension or Schwann cell migration. Intermediate diameter fibers (759 ± 179 nm) promoted long, directed neurite extensions independent of Schwann cell migration, while the large diameter fibers (1325 ± 383 nm) promoted long, directed neurite extension and Schwann cell migration (Wang et al.). NIH 3T3 fibroblasts cultured on electrospun fibers of various diameters showed higher attachment and proliferation rates on small diameter scaffolds (428 versus 1051 nm) (M. Chen et al.). This study also demonstrated that by increasing the fiber diameter to 1647 nm, the proliferation rate could be increased relative to the 1051 nm nanofibers.

In summary, small diameter fibers facilitate cell adhesion and enhance viability as compared to large fiber samples. Initially, on small diameter fibers, more cells will attach and proliferate. At the same time, the associated small interfiber pore size will inhibit cellular infiltration, limiting cells to the scaffold surface with limited contact area for cell spreading. However, large fiber

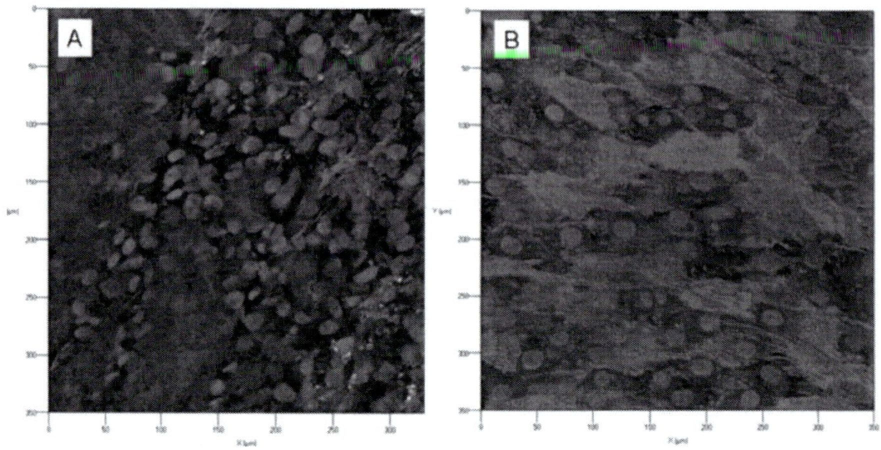

Figure 12. Confocal microscope three-dimensional top view images showing that cells express more F-actin when grown on the scaffold with large fibers. (A) Small diameter fibers (110 nm). (B) Large diameter fibers (600 nm). Red: gelatin autofluorescence of the electrospun scaffold, Green: Phalloidin 488 staining for cytoskeleton (F-actin), Blue: DraQ5 staining for cell nuclei (A) 202 cell nuclei and (B) 115 cell nuclei present in image (Sisson et al. 2010, Reprinted with permission from John Wiley and Sons).

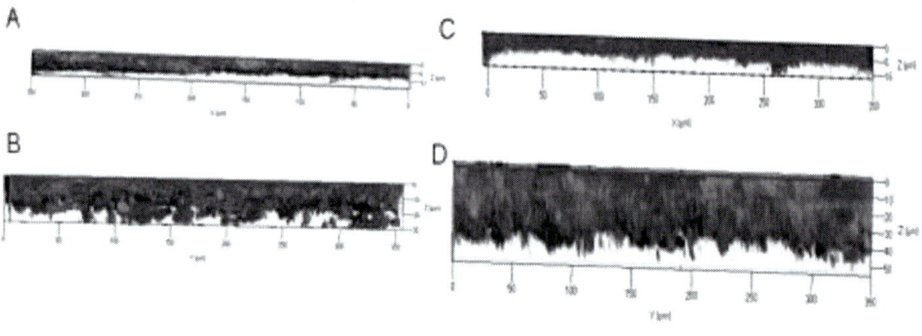

Figure 13. Confocal three-dimensional images illustrate that cells penetrated less into the scaffold with small fiber diameters (A and C) than into those with large fibers (B and D) after 7 or 14 days in culture, respectively. After 7 days in culture, cells have migrated to a maximum depth of 12 μm in the small diameter scaffolds and to a depth of 30 μm in the large diameter scaffolds (A and C). After 14 days in culture, cells have migrated to a maximum depth of 16 μm in the small diameter scaffolds and to a depth of 50 μm in the large diameter scaffolds (B and D). Side view of scaffold with the blue DraQ5 staining for cell nuclei and the green Phalloidin 488 staining for cytoskeleton (F-actin) (Sisson et al. Reprinted with permission from John Wiley and Sons).

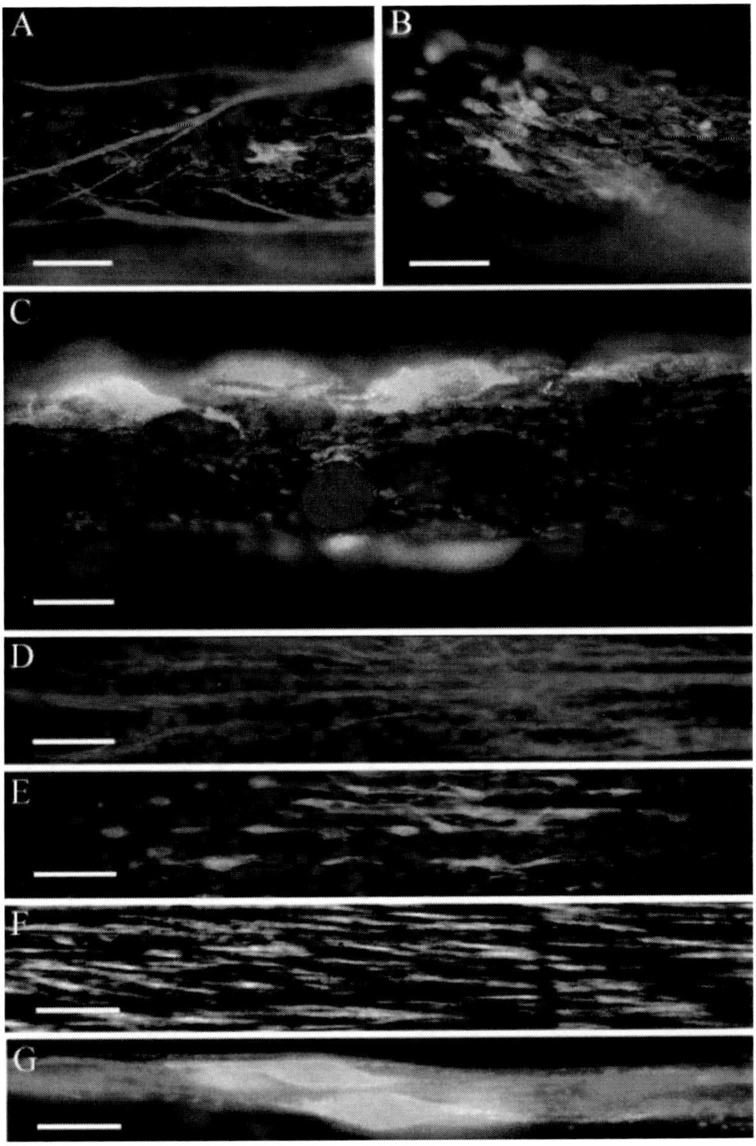

Figure 14. Effect of fiber diameter on neurite outgrowth and alignment. Staining shows neurite outgrowth in red, Schwann cells in green, and cell nuclei in blue. Fiber bundles were arranged horizontally. Bundles are composed of (A) 500 μm diameter fibers, (B) 200 μm diameter fibers, (C) 30 μm diameter fibers, (D-G) 5 μm diameter fibers. Scale bar for all figures except (C) and (G) is 100 μm. In (C) and (G), scale bar is 10 μm. (X. Wen and P. Tresco. Reprinted with permission from John Wiley and Sons).

diameters create relatively larger interfiber pore size which facilitate cell migration into the scaffold (Figure 13) (Sisson et al.). Moreover, cells have the opportunity to touch the surface of interior fibers within the scaffold, potentially generating more surface area for cell spreading. Another interesting observed phenomenon is that for smaller fiber diameters, a single cell will cross a large number of fibers, while on larger fiber diameters a cell will only spread between one to three fibers. This leads to significant differences in cell morphology on different diameter fibers (Figure 14).

2.3. Surface Topology of Fiber

In the last few years, a variety of methods for the fabrication of nano-scale controlled fiber surface topographies have been developed. These methods are important because recent studies have shown that cells are able to respond to this nanopatterning. There are a multitude of ways that cells respond to fiber surface topography including varying their height, width, and shape. In addition, the overall surface organization of these features, the chemistry, and the rigidity of the surfaces as well as the type of cell used in the study will alter the cells' response. Thus, the incorporation of roughness or surface topography on electrospun fibers could be an attractive means to affect biological responses.

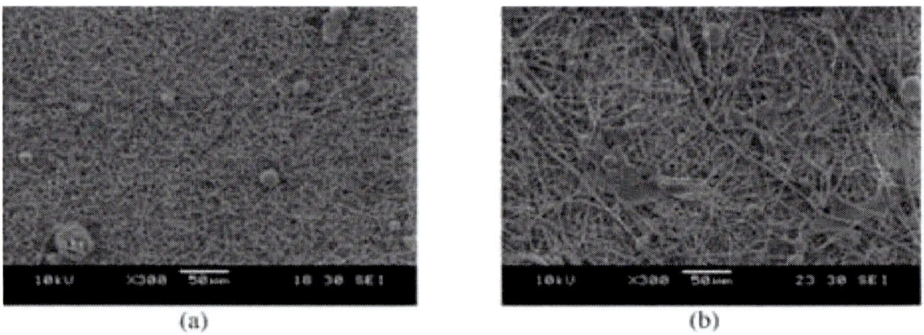

Figure 15. SEM micrographs showing the attachment of porcine esophageal epithelial cells on different electrospun scaffolds. (a) and (b) are micrographs for solid surface and porous fiber scaffolds for 2 h, respectively (Leong et al. Reprinted with permission from John Wiley and Sons).

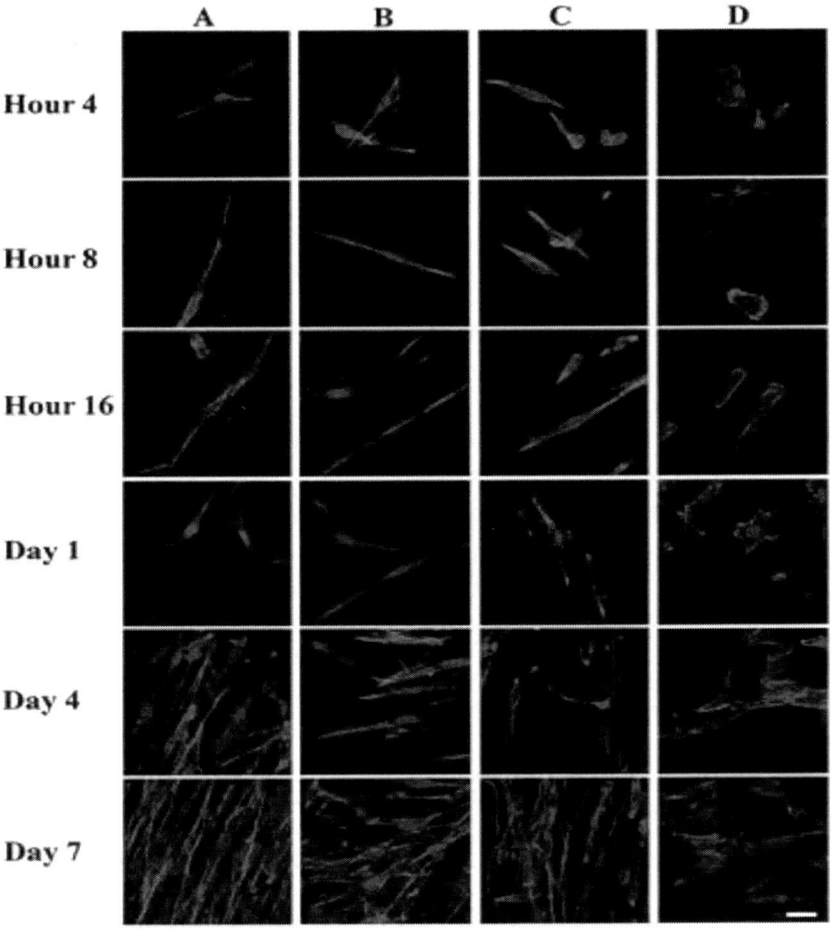

Figure 16. Typical confocal microscopic images of Schwann cells grown on (A) aligned cellulose acetate butyrate nanofibers having a parallel line surface texture, (B) randomly oriented cellulose acetate butyrate nanofibers having a parallel line surface texture, (C) aligned cellulose acetate butyrate nanofibers having a smooth surface, and (D) randomly oriented cellulose acetate butyrate nanofibers having a smooth surface, after different culture periods. Scale bar equals 25 μm (Huang et al. Reprinted with permission from the Royal Society of Chemistry).

The most commonly fabricated and tested textures on electrospun fiber surfaces are nanoporosity, irregular roughness, and parallel stripes. The fiber surface nano-pores, ranging from 50 nm to 500 nm, can be systematically controlled by adjusting solution concentration, solvent volatility, and relative humidity (Cao Shengguang; Kongkhlang et al., Kim et al., Megelski et al.). As an example, Leong et al. introduced nano-pores into the surface of electrospun poly(L-lactide acid) fibers and showed that the porous surface leads to a significant increase in the amount of protein adsorption (1.5 times), which promotes cell attachment. Porcine esophageal epithelial cells seeded onto these nano-porous poly(L-lactide acid) fibers adhered faster than those seeded onto solid fibers (Figure 15).

From these results, they hypothesized that the increase in surface area of the nano-porous fibers provides additional anchorage points for the cells that, along with the increased protein adsorbed, promoted cell attachment. Corona plasma treatment has also been utilized to modify the nano-scale surface properties of electrospun fibers (Nandakumar et al.). It was found that the plasma-modified surface properties showed no effect on the attachment and proliferation of human mesenchymal stem cells, but had a positive effect on the expression of osteogenic genes osteonectin (ON) and bone sialoprotein (BSP) after 7 days of culture and on the activity of alkaline phosphatase after 21 days. Cellulose acetate butyrate nanofibers with parallel stripes have been fabricated using a solvent mixture of acetone and N,N'-dimethylacetamide (Huang et al.). *In vitro* experiments indicated that cells on the nanofibers with a parallel line surface structure showed a slightly better alignment. This finding indicates that parallel line surface structures could facilitate cell extension at the early stage of cell culture (Figure 16).

2.4. Interfiber Pore Size

The natural extracellular matrix has a three-dimensional structure. To mimic the structure of the extracellular matrix, three-dimensional scaffolds are required. Cellular infiltration into these scaffolds is essential for a range of tissue engineering applications and is particularly important in regenerating three-dimensional tissues such as skin and muscle. The scaffold pore size, porosity, and pore interconnectivity dictate the extent of cellular infiltration and tissue ingrowth, influence a range of cellular processes, and are crucial for diffusion of nutrients, metabolites, and waste products (Rnjak-Kovacina and Weiss). In general, electrospun fibers have a small diameter and result in densely packed structures, with small pores between adjacent fibers that can often inhibit the penetration of cells into the scaffolds. Larger pores tend to favor the migration and infiltration of cells. To facilitate infiltration while retaining extracellular matrix-like character, electrospinning has been combined with several methods to produce fine diameter three-dimensional scaffolds with enhanced porosity. Some of these methods include the introduction of porogens, i.e., salt leaching (Hong and Madihally), yarn formation (Wu et al.), layered scaffold (C. Vaquette and J. Cooper-White), water bath collecting (Yeatts et al.), tailored collectors (Blakeney et al., C. Vaquette and J. J. Cooper-White), and air-impedance electrospinning (McClure et al.).

Ki et al. used a coagulation bath containing methanol as a collector to create electrospun silk scaffolds exhibiting a highly porous structure, which promoted the infiltration of cells into the scaffold (Figure 17). This method allowed fibers to disperse uniformly throughout the coagulation bath during electrospinning, creating a suspension. The methanol was then exchanged for water using a cellulose membrane, and the structure was then lyophilized to preserve its architecture. Mouse fibroblasts showed excellent adhesion and proliferation on both conventionally spun and methanol-collected scaffolds, but the methanol-collected scaffolds showed much higher cell infiltration than the conventionally spun scaffolds (Ki et al.).

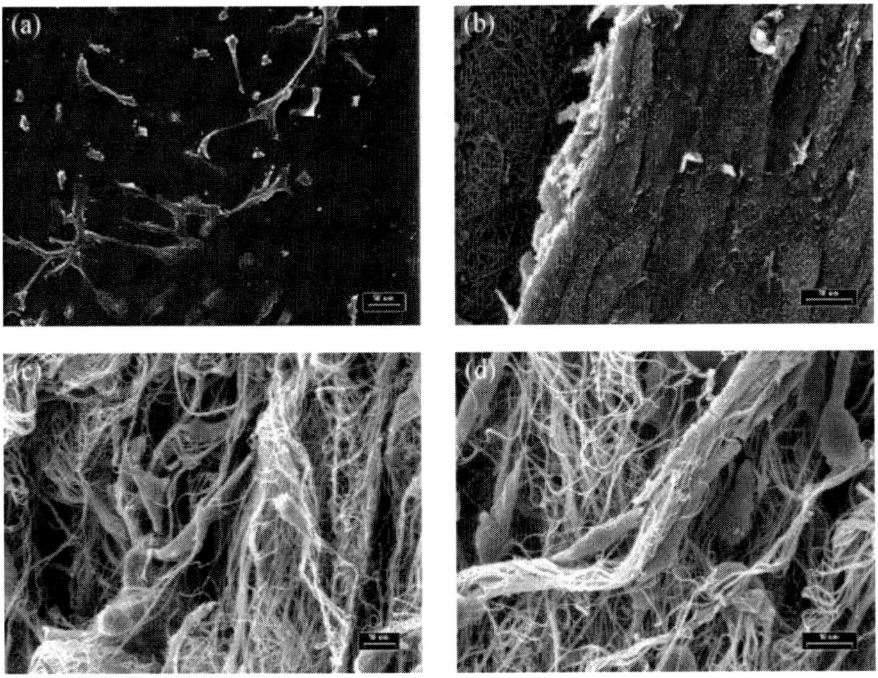

Figure 17. Scanning electron microscopy images of mouse fibroblasts cultured on (a) silk fibroin film, (b) conventionally spun silk fibroin nanofiber mat, (c,d) methanol-collected silk fibroin nanofiber structures. (Ki et al. Reprinted with permission from John Wiley and Sons).

Another novel method of increasing scaffold porosity, developed by Thorvaldsson et al., involves electrospinning nanofibers onto a previously electrospun microfiber in order to create nanofiber-coated microfibers. The microfiber is placed in the direction of electrospinning, with one end by the needle of the syringe and the other end placed along the edge of a rotating disk, and is surrounded by a Teflon tube to ensure proper collection of nanofibers. Nanofibers are then spun into the Teflon tube, where they land on the microfiber as it rotates within the tube, ensuring a full 360 degree coating of nanofibers (Figure 18). This method was used to create polycaprolactone microfibers coated with polycaprolactone nanofibers, and these coated fibers were then pressed into three-dimensional scaffolds.

Chondrocytes cultured on the resulting scaffolds grew along the microfibers, reaching out and connecting to additional nanofibers at the interconnections, and showed much better infiltration into these scaffolds than into the pure nanofiber control scaffolds. In control scaffolds of non-coated microfibers, the chondrocytes were able to infiltrate the scaffold, but formed sheet structures with less adherence to the fibers (Figure 19) (Thorvaldsson et al.). Thus, the nanofiber-coated microfiber scaffolds combined the promotion of cellular adherence of the nanofiber scaffolds with the cellular infiltration of the microfiber scaffolds to create more effective tissue engineering scaffolds.

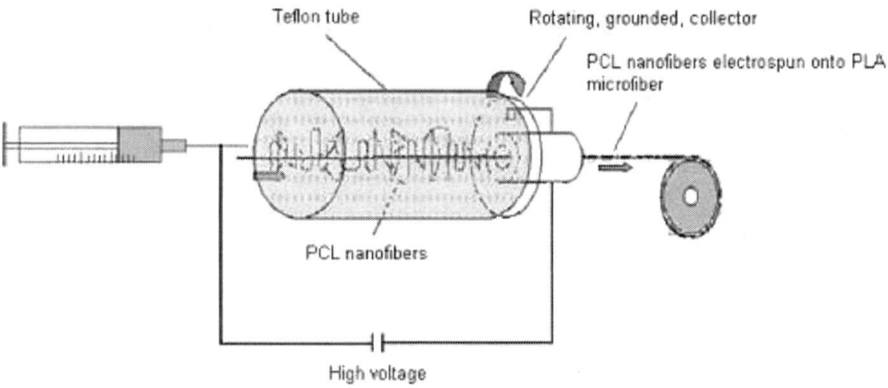

Figure 18. Method of electrospinning nanofibers onto a microfiber. The microfiber is placed along the electrospinning axis and rotated to allow for complete nanofiber surrounding. (Thorvaldsson et al. Reprinted with permission from the American Chemical Society).

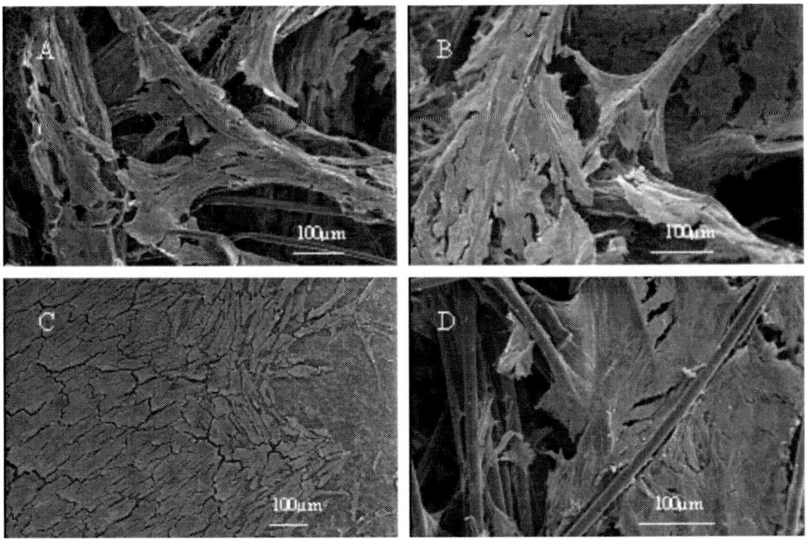

Figure 19. Chondrocytes cultured on PLA microfibers coated with polycaprolactone nanofibers. Scaffolds pictured have (A) 97% and (B) 95% porosity. Referene pictures of (C) uncoated polycaprolactone nanofibers and (D) PLA microfibers only. (Thorvaldsson et al. Reprinted with permission from the American Chemical Society).

Vaquette et al. utilized patterned collectors to tailor the morphology of associated electrospun scaffolds. The results demonstrate that circular macropore patterning leads to an increased pore size for colonization and penetration of cells. Electrospun fiber yarn fabricated via a dynamic liquid (water vortex) collection method generated large pores by the formation of fiber bundles. Wu et al. illustrated that these loosely packed structures could significantly facilitate cell proliferation. On the other hand, electrospun scaffolds with enhanced pore sizes always show a lower regional fiber density, which reduces both tensile strength and compressive strength. Thus, regardless of method to enhance scaffold pore size and porosity, it will be critical to maintain the mechanical properties for handling, suturing, and bulk tissue regeneration.

2.5. Incorporating Cells during the Electrospinning Process

One potential way to increase cellular incorporation in an electrospun scaffold without sacrificing the mechanical integrity of that scaffold is to incorporate those cells directly during the electrospinning process, rather than inducing their infiltration after scaffold formation. This technique, first proposed by Sanders et al. in 2003, has been studied more extensively in recent years and several research groups have succeeded in electrospinning or electrospraying suspensions of live cells. Townsend-Nicholson et al. developed a coaxial bioelectrospraying system that successfully incorporated a suspension of live 1321N1 astrocytoma cells within electrospun fibers (Figure 20).

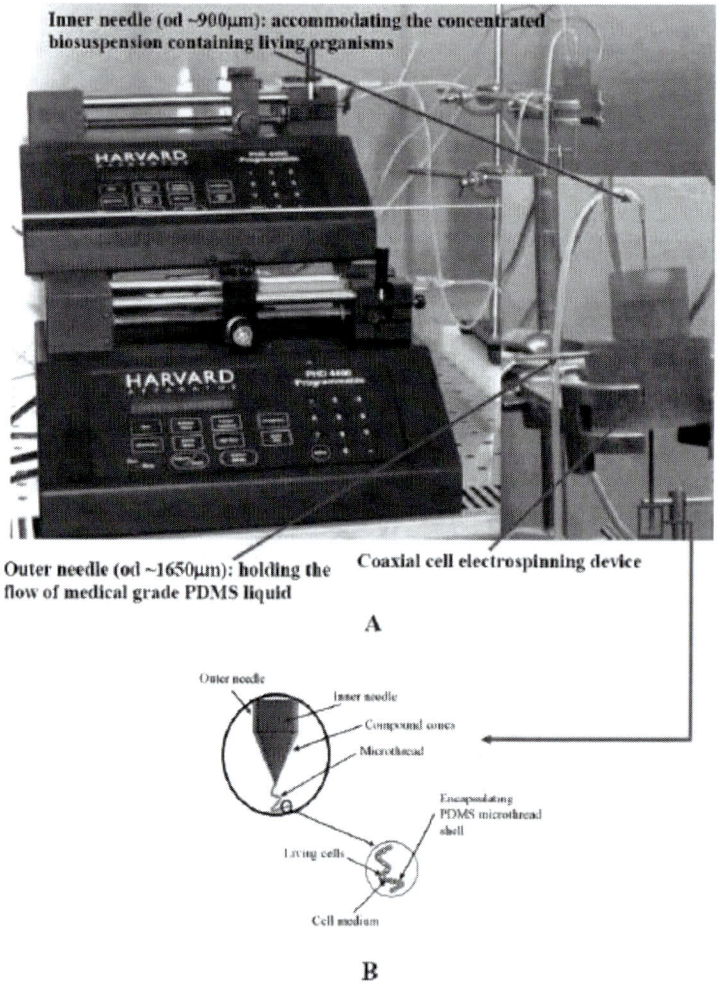

Figure 20. (A) Coaxial electrospinning device setup. (B) Schematic of the needle tip, showing the process of encapsulating the cell suspension within the electrospun fiber. (Townsend-Nicholson et al. Reprinted with permission of the American Chemical Society)

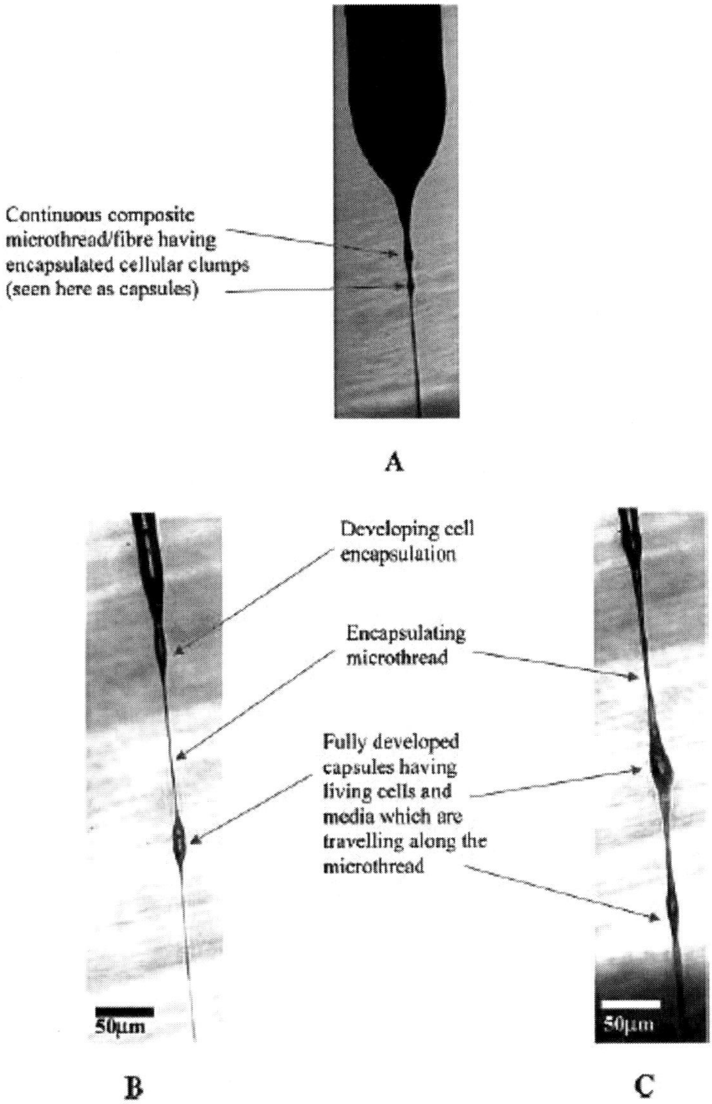

Figure 21. (A) Close up, high speed image of electrospinning showing capsules of cell clumps as they exit the needle tip within a fiber. (B) Spinning of fine fibers with small cell encapsulations. (C) Spinning of larger diameter fibers with larger cell encapsulations (Townsend-Nicholson et al. Reprinted with permission of the American Chemical Society).

This system extruded cell suspension through an inner needle while medical-grade polydimethylsiloxane was extruded from a surrounding needle, creating a polymer shell around the suspension (Figure 21). The resulting electrospun fibers incorporated cell-containing capsules, creating a beads-on-a-string morphology. Parameters were optimized to create finer threads with small diameter encapsulations or thicker threads with larger cell encapsulations. A cell viability assay determined that there was no observable impact of the electrospinning process on overall cell viability up to 7 days. Imaging and analysis demonstrated no observable impact on cell morphology due to the electrospinning process itself, although the polydimethylsiloxane fiber shell caused chainlike structures in the final cell collection. Nevertheless, the procedure laid out in this study is a simple and effective way to incorporate cells within an electrospun scaffold without a loss in viability.

In a similar study, Ang et al. electrospun human umbilican vein endothelial cells (HUVECs) into a polyethylene oxide core/polycaprolactone porous shell fiber to create bioactive fibers for the delivery of therapeutic biomolecules released by the cells. A coaxial electrospinning setup was used, in which the inner cell/polyethylene oxide core nanofiber was surrounded by a coat of polycaprolactone upon emission from the apparatus to create core/shell fibers. Fibers were electrospun into a water collector in order to impart the outer shell with pore sizes of 1.7 ±.4 μm, with the aim of promoting the exchange of nutrients and waste between the cells and the outside environment as well as promoting the release of biomolecules from the cells. Fetal Bovine Serum (FBS) was incorporated into the PEO core of the fibers along with the HUVECs in order to improve cell viability, and enabled the cells to survive up at 2 weeks post-electrospinning. An Interleukin-8 (IL-8) ELISA showed the continuous release of IL-8 from the nanofibers, indicating sufficient fiber shell porosity. While initial levels of IL-8 release from entrapped HUVECs was higher than non-entrapped control, indicating shear stress on these cells, IL-8 release returned to constitutive levels after four days indicating the alleviation of that stress (Ang et al.). This method of cell encapsulation during the electrospinning process shows great promise as a way to deliver prolonged levels of bioactive molecules through the secretions of entrapped cells.

Sampson et al. compared cell electrospinning to a similar method of encapsulating cells within nanofibers, called bio-threading. In this method, a pressure drop is used to pull polymer solution through an orifice to produce continuous polymer fibers, which are collected into a scaffold. The main advantage of this method is that the cell suspension does not have to be electrically conductive, and the cells are not subject to the electrical charges used in electrospinning. Scaffolds with encapsulated mouse neuroblastoma cells within the fibers were created by both cell electrospinning and bio-threading (Figure 22).

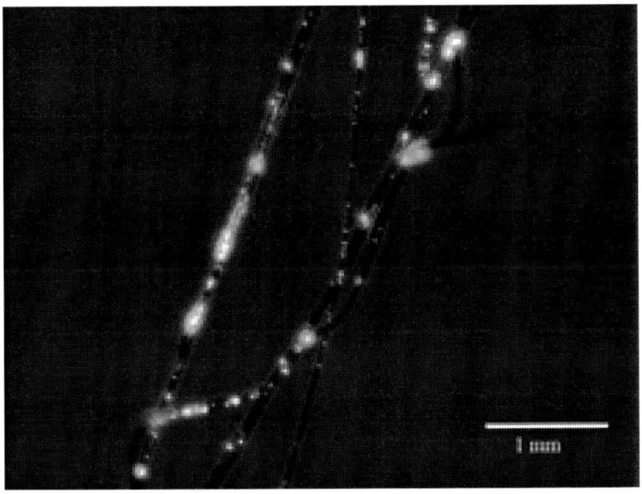

Figure 22. Cell-encapsulating fibers created by electrospinning a blend of cells and polymer. Similar fibers are made by way of the bio-threading procedure. (Sampson et al. Reprinted with permission from John Wiley and Sons)

Flow cytometry showed good cell viability maintained in both methods. Scaffold constructs were then implanted into mouse abdomens and left to grow *in vivo* for 13 days (Figure 23). Subsequent analysis showed no detectable effect of either the cell electrospinning or the biothreading on the *in vitro* or *in vivo* growth of the cells. While the main aim of the study was to ensure that these processes did not adversely impact the cells, the mouse abdomen implantation demonstrated that both types of composite cellularized scaffolds integrated into the host without rejection during the 13 day implantation period (Sampson et al.).

Hong et al. used electrospraying, a technique similar to electrospinning, to encapsulate cells within polydimethylsiloxane nanoparticles. In this study, a microfluidic device was used to create droplets of Jurkat cells suspended in an alginate solution. These droplets were exposed to an oil-phase polydimethylsiloxane solution and then extruded through a needle at high voltage at a grounded electrode. Instead of forming fibers like an electrospun solution, these droplets formed nanoparticles with a cell-containing alginate core and a polydimethylsiloxane outer layer. They were able to vary the number of cells per droplet by controlling the ratio of aqueous flow to oil flow, creating larger or smaller droplets. An assessment of post-cell encapsulation viability showed cell mitosis to occur on the same timetable as control cells, indicating no adverse effects of the encapsulation technique on the cells. An image of a typical setup used to electrospray cells is shown in Figure 24.

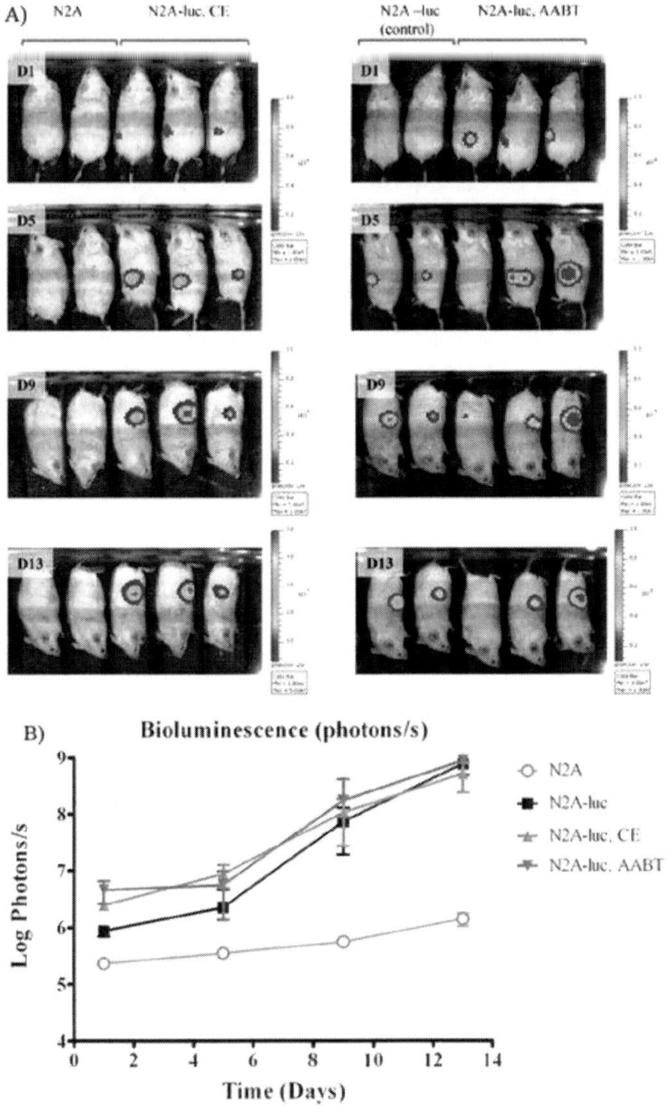

Figure 23. Bioluminescent images of mice implanted with (A) 5000 untreated neuroblastoma cells (labeled N2A), untreated luciferase-expressing neuroblastoma cells (labeled N2A-Luc), cell-electrospun luciferase-expressing neuroblastoma cells (labeled N2A-Luc, CE), and bio-threaded luciferase-expressing neuroblastoma cells (N2A-Luc, AABT). Mice were imaged at 1, 5, 9, and 13 days after implantation. Images are displayed as pseudocolor images of peak bioluminescence. (B) Quantification of bioluminescence. A one-way ANOVA with Bonferroni post-hoc showed no significant differences. (Sampson et al. Reprinted with permission from John Wiley and Sons).

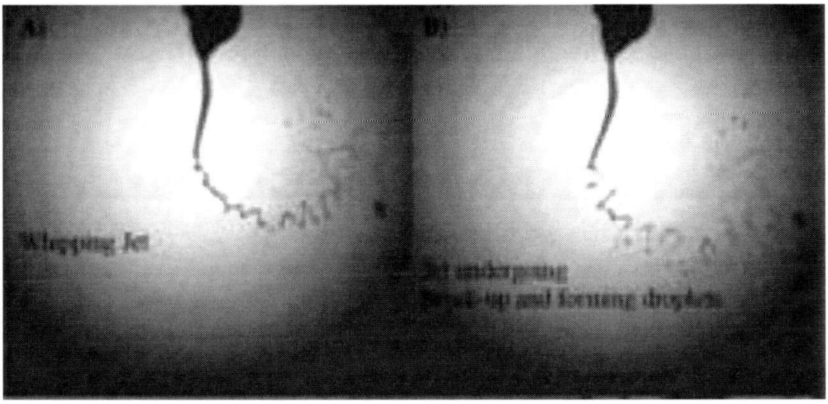

Figure 24. Images of a typical bioelectrospraying setup. Polymer/cell solution is emitted under conditions that promote an unstable whipping jet. As the jet breaks up, cell-encapsulating polymer particles are formed. (Andreu et al. Reprinted with permission of John Wiley and Sons).

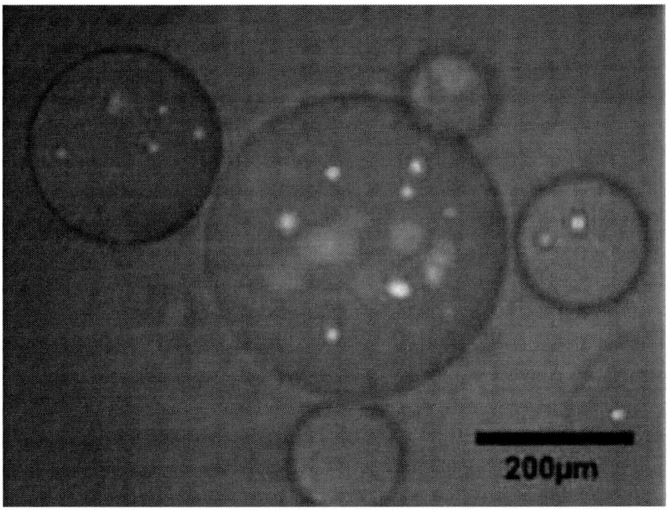

Figure 25. Fluorescently labeled mouse fibroblasts encapsulated in poly(vinyl alcohol) microspheres 24 hours post-electrospraying. Live cells are shown as green, while dead cells are shown as red. (Young, Poole-Warren and Martens. Reprinted with permission from John Wiley and Sons.).

In a similar study, Young, Poole-Warren, and Martens electrosprayed mouse fibroblast-laden poly(vinyl alcohol) PVA microspheres using a technique called submerged electrospraying, in which the polymer beads are formed in an insulating liquid instead of in air. A syringe pumped the cell/poly(vinyl alcohol) solution through a high-voltage needle submerged in sunflower oil with an earthed ring at the bottom of the beaker. As the solution emerged from the needle tip, it broke up into microbeads which were then crosslinked by UV light as they flew through the oil towards the grounded copper ring. By varying typical electrospraying parameters such as flow rate, voltage, and distance, microspheres ranging from around 10 μm to 3 mm diameters were created. Cells were fluorescently stained to assess encapsulation and viability, and were found to be over 95% viable and maintain their ability to differentiate post-encapsulation (Figure 25) (Young, Poole-Warren, and Martens).

A study by Stankus et al. combined both electrospinning and electrospraying to incorporate smooth muscle cells into an electrospun vascular graft. Their setup involved the use of two separate nozzles, one electrospinning poly(ester urethane) urea fibers and the other electrospraying a solution of vascular smooth muscle cells, creating a scaffold with cells evenly interspersed throughout the nanofiber matrix.

Unlike the method used by Hong et al., the cells were not encapsulated within the fibers but distributed between them, more closely mimicking the location of cells in natural extracellular matrix. A high mandrel rotation speed relative to the translational speed was used in order to create scaffolds with the ability to withstand high circumfrential stresses. Scaffolds were then placed in static culture for one day, randomized, and then placed in either static culture or in spinner flask culture at 15 rpm. An MTT mitochondrial activity assay showed high numbers of viable cells, and imaging of H&E-stained sections corroborated this finding. Mechanical testing indicated that these constructs maintained a high burst pressure, ultimate tensile stress, and a high distensibility, giving them similar mechanical properties to those of healthy coronary arteries. H&E stained sections demonstrated that the spinner flask-cultured scaffolds had improved cell number and better cell spreading than the statically cultured samples, indicating better nutrient availability and waste removal under these experimental conditions.

References

Andreu et al. "In Vitro and In Vivo Interrogation of Bio-sprayed Cells." *Small* 8 (2012): 2495-2500.

Ang, H.Y. et al. "Characterization of a Bioactive Fiber Scaffold with Entrapped HUVECs in Coaxial Electrospun Core-Shell Fiber." *Biomatter* 4 (2014): e28238-1 – e28238-7.

Ayres, C.E., Shekhar Jha, B., Meredith, H., Bowman, J.R., Bowlin, G.L., Henderson, S.C. and D.G. Simpson. "Measuring Fiber Alignment in Electrospun Scaffolds: A User's Guide to the 2D Fast Fourier Transform Approach." *Journal Biomaterials Science: Polymer Edition* (2008) 19: 603-21.

Baker, B.M., et al. "The Potential to Improve Cell Infiltration in Composite Fiber-Aligned Electrospun Scaffolds by the Selective Removal of Sacrificial Fibers." *Biomaterials* 29 (2008): 2348-58.

Bashur, C.A., L.A. Dahlgren, and A.S. Goldstein. "Effect of Fiber Diameter and Orientation on Fibroblast Morphology and Proliferation on Electrospun Poly(D,L-Lactic-Co-Glycolic Acid) Meshes." *Biomaterials* 27 (2006): 5681-8.

Bashur, Chris A., et al. "Effect of Fiber Diameter and Alignment of Electrospun Polyurethane Meshes on Mesenchymal Progenitor Cells." *Tissue Engineering Part A* 15 (2009): 2435-45.

Blakeney, B.A., et al. "Cell Infiltration and Growth in a Low Density, Uncompressed Three-Dimensional Electrospun Nanofibrous Scaffold." *Biomaterials* 32 (2011): 1583-90.

Shengguang, C., Binghuan, H. and L. Haiqing. "Fabrication of Nano-Porous Structured Polylactide (PLLA) Fibers through Electrospinning." *Acta Polymerica Sinica* 10 (2010): 1193-98.

Chen, M., et al. "Role of Fiber Diameter in Adhesion and Proliferation of NIH 3T3 Fibroblast on Electrospun Polycaprolactone Scaffolds." *Tissue Engineeering* 13 (2007): 579-87.

Chew, S.Y., et al. "The Effect of the Alignment of Electrospun Fibrous Scaffolds on Schwann Cell Maturation." *Biomaterials* 29 (2008): 653-61.

Choi, J.S., et al. "The Influence of Electrospun Aligned Poly(Epsilon-Caprolactone)/Collagen Nanofiber Meshes on the Formation of Self-Aligned Skeletal Muscle Myotubes." *Biomaterials* 29 (2008): 2899-906.

Christopherson, G.T., H. Song, and H.Q. Mao. "The Influence of Fiber Diameter of Electrospun Substrates on Neural Stem Cell Differentiation and Proliferation." *Biomaterials* 30 (2009): 556-64.

Corey, J. M., et al. "Aligned electrospun nanofibers specify the direction of dorsal root ganglia neurite growth." *Journal of Biomedical Materials Research* 83 (2007): 636-645.

Discher, D.E., Janmey, P. and Y.L. Wang. "Tissue Cells Feel and Respond to the Stiffness of Their Substrate." *Science* 310 (2005): 1139-43.

Erisken, C., et al. "Scaffold Fiber Diameter Regulates Human Tendon Fibroblast Growth and Differentiation." *Tissue Engineering Part A* 19 (2013): 519-28.

He, L., et al. "Synergistic Effects of Electrospun PLLA Fiber Dimension and Pattern on Neonatal Mouse Cerebellum C17.2 Stem Cells." *Acta Biomaterialia* 6 (2010): 2960-9.

Hong, J.K., and S.V. Madihally. "Three-Dimensional Scaffold of Electrosprayed Fibers with Large Pore Size for Tissue Regeneration." *Acta Biomaterialia* 6 (2010): 4734-42.

Hong, J.K., et al. "Bio-electrospraying and Droplet-Based Microfluidics: Control of Cell Numbers Within Living Residues."*Biomedical Materials* 5 (2010): 021001.

Huang, C., et al. "Electrospinning of Nanofibres with Parallel Line Surface Texture for Improvement of Nerve Cell Growth." *Soft Matter* 7 (2011): 10812.

Ingber, D.E. "Mechanical Signaling and the Cellular Response to Extracellular Matrix in Angiogenesis and Cardiovascular Physiology." *Circulation Research* 91 (2002): 877-87.

Kai, D., et al. "Guided Orientation of Cardiomyocytes on Electrospun Aligned Nanofibers for Cardiac Tissue Engineering." *Journal of Biomedical Materials Research Part B* 98 (2011): 379-86.

Ki, C. S., et al. "Electrospun Three-Dimensional Silk Fibroin Nanofibrous Scaffold." *Journal of Applied Polymer Science* 106 (2007): 3922-3928.

Kim, B.S. and D.J. Mooney. "Development of Biocompatible Synthetic Extracellular Matrices for Tissue Engineering." *Trends in Biotechnology* 16 (1998): 224-30.

Kim, C.H., et al. "Effect of Collector Temperature on the Porous Structure of Electrospun Fibers." *Macromolecular Research* 14 (2006): 59-65.

Kongkhlang, T., et al. "Electrospun Polyoxymethylene: Spinning Conditions and Its Consequent Nanoporous Nanofiber." *Macromolecules* 41 (2008): 4746-52.

Li, H., et al. "Electrospun Membranes: Control of the Structure and Structure Related Applications in Tissue Regeneration and Drug Delivery." *Journal of Materials Chemistry B* (2014): 5492–5510.

Lee, C.H., et al. "Nanofiber Alignment and Direction of Mechanical Strain Affect the ECM Production of Human ACL Fibroblast." *Biomaterials* 26 (2005): 1261-70.

Leong, M.F., et al. "Effect of Electrospun Poly (D, L-Lactide) Fibrous Scaffold with Nanoporous Surface on Attachment of Porcine Esophageal Epithelial Cells and Protein Adsorption." *Journal of Biomedical Materials Research Part A* 89 (2009): 1040-48.

Martins-Green, M. "The Dynamics of Cell-ECM Interactions with Implications for Tissue Engineering." In: Principles of Tissue Engineering, ed., Lanza, R., Langer, R., and Chick, W., editor. R. G. Landes Company (1997): 23-46.

McClure, M.J., et al. "The Use of Air-flow Impedance to Control Fiber Deposition Patterns during Electrospinning." *Biomaterials* 33 (2012): 771-779.

Megelski, S., et al. "Micro-and Nanostructured Surface Morphology on Electrospun Polymer Fibers." *Macromolecules* 35 (2002): 8456-66.

Muzzarelli, R.A., et al. "Chitosan, Hyaluronan and Chondroitin Sulfate in Tissue Engineering for Cartilage Regeneration: A Review." *Carbohydrate Polymers* 89 (2012): 723-39.

Nandakumar, A., et al. "Surface Modification of Electrospun Fibre Meshes by Oxygen Plasma for Bone Regeneration." *Biofabrication* 5 (2013): 015006.

Palsson, B.O. and Bhatia, S.N., Tissue Engineering, Pearson Prentice Hall, Upper Sadle River, New Jersey; (2004).

Patch, K, and E.B. Smalley. "Handles Nanofiber Better." *Technology Research News* 17 (2003): 24.

Pattison, M.A., et al. "Three-dimensional, Nano-structured PLGA Scaffolds for Bladder Tissue Replacement Applications." *Biomaterials* 26 (2005): 2491–2500.

Piez, K.A. "Molecular and Aggregate Structures of the Collagens." Elsevier, New York, 1984. Vol. 5.

Ramakrishna, S., et al. *An Introduction to Electrospinning and Nanofibers*. Vol. 90: World Scientific, 2005.

Rnjak-Kovacina, J., and A.S. Weiss. "Increasing the Pore Size of Electrospun Scaffolds." *Tissue Engineering Part B* 17 (2011): 365-72.

Sampson, S. et al. "Cell Electrospinning: An *In Vitro* and *In Vivo* Study." *Small* 10(2014): 78-82.

Sanders, J., Lamont, S.E., Karchin, A., Golledge, S.L., Ratner, B.D. "Fibro-porous Meshes Made from Polyurethane Micro-fibers: Effects of Surface Charge on Tissue Response." *Biomaterials* 26: 813-818 (2005).

Sanders, E. H., et al. "Two-Phase Electrospinning from a Single Electrified Jet: Microencapsulation of Aqueous Reservoirs in Poly(ethylene-co-vinyl acetate) Fibers." *Macromolecules* 36 (2003): 3803-3805.

Sangsanoh, P., et al. "*In vitro* Biocompatibility of Schwann Cells on Surfaces of Biocompatible Polymeric Electrospun Fibrous and Solution-Cast Film Scaffolds." *Biomacromolecules* 8 (2007): 1587-94.

Sell, S.A., McClure, M.J., Ayers, C.E., Simpson, and G.L. Bowlin. "Preliminary Investigation of Airgap Electrospun Silk Fibroin-based Structures for

Ligament Analogue Engineering." *Journal of Biomaterials Science: Polymer Edition* 22 (2011): 1253-1273.

Shih, Y.R., et al. "Growth of Mesenchymal Stem Cells on Electrospun Type I Collagen Nanofibers." *Stem Cells* 24 (2006): 2391-2397.

Simsek, M., et al. "Chitosan and Polycaprolactone Membranes Patterned Via Electrospinning: Effect of Underlying Chemistry and Pattern Characteristics on Epithelial/Fibroblastic Cell Behavior." *Journal of Biomedical Materials Research Part A* 100 (2012): 3332-43.

Sisson, K., et al. "Fiber Diameters Control Osteoblastic Cell Migration and Differentiation in Electrospun Gelatin." *Journal of Biomedical Materials Research Part A* 94 (2010): 1312-20.

Stankus, J, et al. "Fabrication of Cell Microintegrated Blood Vessel Constructs Through Electrohydrodynamic Atomization." *Biomaterials* 28 (2007): 2738-2746.

Thompson, C.J., et al. "Effects of Parameters on Nanofiber Diameter Determined from Electrospinning Model." *Polymer* 48 (2007): 6913-22.

Thorvaldsson, A., et al. "Electrospinning of Highly Porous Scaffolds for Cartilage Regeneration." *Biomacromolecules* 9 (2008): 1044-1049.

Townsend-Nicholson, A., and S.N. Jayasinghe. "Cell Electrospinning: a Unique Biotechnique for Encapsulating Living Organisms for Generating Active Biological Microthreads/Scaffolds." *Biomacromolecules* 7 (2006): 3364-3369.

Vaquette, C., and J. Cooper-White. "A Simple Method for Fabricating 3-D Multilayered Composite Scaffolds." *Acta Biomaterialia* 9 (2013): 4599-608.

Vaquette, C., and J.J. Cooper-White. "Increasing Electrospun Scaffold Pore Size with Tailored Collectors for Improved Cell Penetration." *Acta Biomaterialia* 7 (2011): 2544-57.

Wang, H.B., et al. "Varying the Diameter of Aligned Electrospun Fibers Alters Neurite Outgrowth and Schwann Cell Migration." *Acta Biomaterialia* 6 (2010): 2970-8.

Wen, X., and P.A. Tresco. "Effect of Filament Diameter and Extracellular Matrix Molecule Precoating on Neurite Outgrowth and Schwann Cell Behavior on Multifilament Entubulation Bridging Device *In Vitro*." *Journal of Biomedical Materials Research* 76 (2006): 626-637.

Wu, J., et al. "Electrospun Nanoyarn Scaffold and Its Application in Tissue Engineering." *Materials Letters* 89 (2012): 146-49.

Yeatts, A.B., et al. "*In vivo* Bone Regeneration Using Tubular Perfusion System Bioreactor Cultured Nanofibrous Scaffolds." *Tissue Engineering Part A* 20 (2014): 139-46.

Yin, Z., et al. "The Regulation of Tendon Stem Cell Differentiation by the Alignment of Nanofibers." *Biomaterials* 31 (2010): 2163-75.

Young, C.J., Poole-Warren, L.A. and P.J. Martens. "Combining Submerged Electrospray and UV Photopolymerization for Production of Synthetic Hydrogel Microspheres for Cell Encapsulation." *Biotechnology and Bioengineering* 109 (2012): 1561-1570.

Chapter 3

APPLICATION OF ELECTROSPUN NANOFIBERS IN TISSUE ENGINEERING

ABSTRACT

The fabrication of tissue engineering scaffolds through electrospinning allows the fibrous nature and composition of the native extracellular matrix to be mimicked within the scaffold, making electrospinning an extremely popular technique for potential use in regenerating tissue defects. Moreover, electrospun scaffolds can be easily manufactured with diverse morphologies, facilitating their use in a variety of applications. In this chapter, we review several popular applications of electrospun scaffolds which demonstrate their potential in the field of tissue engineering. We hope that this review will inspire new ideas promoting the development of the next generation of electrospun scaffolds and the eventual creation of functional tissue engineered products.

Keywords: electrospinning, tissue engineering, blood vessel, bone, cartilage, ligament, skin

INTRODUCTION

Biomimetic electrospun fibers represent a new frontier of innovative scaffold fabrication methods for applications such as nerve conduits, small diameter arteries, and skin. These methods raise tissue engineering to a new stage, namely the rapid, flexible production of biomimetic extracellular matrices. The flexibility is derived from the fact that almost all the biomaterials normally used in scaffold fabrication can be processed into electrospun fibrous scaffolds. These materials include synthetic polymers such as poly(lactic acid), poly(glycolic acid), polycaprolactone and their copolymers, as well as a large variety of natural polymers such as collagen, gelatin, silk, spider silk, hyaluronic acid, chitosan, and cellulose (Min et al., H. Yoshimoto et al.). Along with the flexibility of the source material, electrospinning offers the ability to produce a variety of different structures by altering their processing parameters (Chapter 2). These alterations include making micro- to nano-scale fibers, varying the degree of fiber alignment, arranging layers in random, parallel, or crossing structures, or macro- and micro-patterning the fibers. The fiber orientation allows for the regulation of cellular alignment and induced morphology as well as overall tissue biomechanics. Due to the preferential alignment of cells in the direction of the electrospun fiber orientation (C. Xu et al.), these scaffolds can be designed for the regeneration of tissues with highly oriented architectures, i.e., nerves and tendons.

Another aspect of electrospun fibrous scaffold processing, beyond the simulation of the structure of the extracellular matrix (Figure 26),

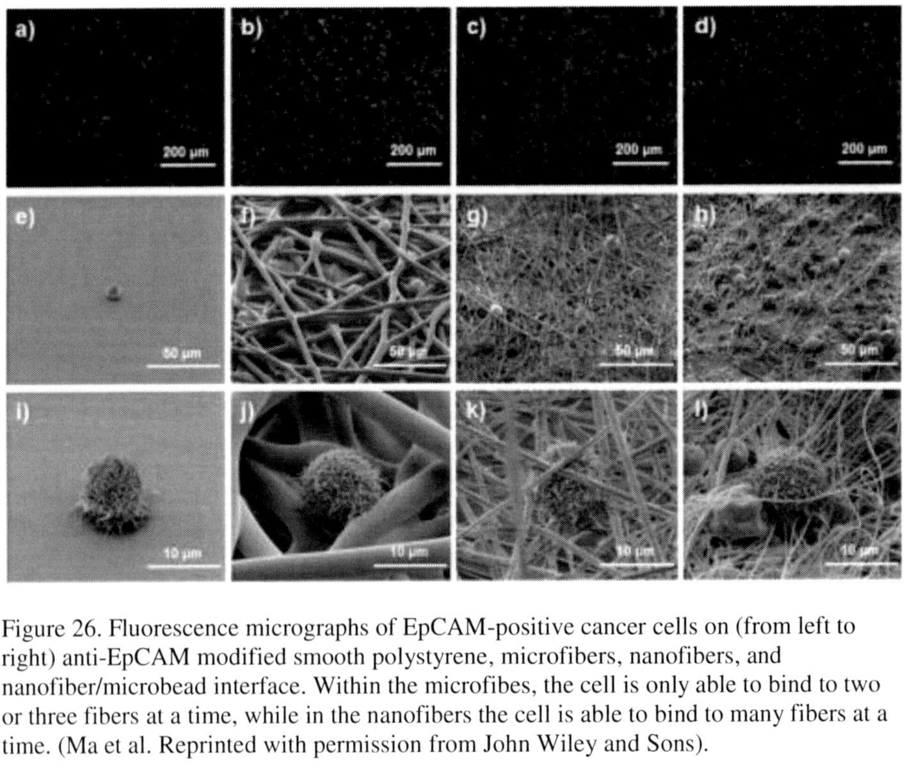

Figure 26. Fluorescence micrographs of EpCAM-positive cancer cells on (from left to right) anti-EpCAM modified smooth polystyrene, microfibers, nanofibers, and nanofiber/microbead interface. Within the microfibes, the cell is only able to bind to two or three fibers at a time, while in the nanofibers the cell is able to bind to many fibers at a time. (Ma et al. Reprinted with permission from John Wiley and Sons).

is the development of functionalized fibers by coaxial electrospinning. Coaxial electrospinning allows one to add within the fibers various functional factors (i.e., growth factors) in order to use the fibers as a controlled release vehicle. This gives the tissue engineering scaffolds an additional level of functionality in the development of tissue constructs. One example of this technique is the encapsulation of heparin in the electrospun fibers of a vascular regeneration scaffold encompassing a stent to provide anticoagulant protection during healing (Su et al.). Other examples include electrospinning nerve growth factor into fibrous nerve conduits to promote nerve regeneration (C. Y. Wang et al.), and incorporating bone morphogenetic protein within electrospun fibrous scaffolds to induce the differentiation of mesenchymal stem cells into osteoblasts to form new bone tissue. Therefore, many believe that this technique may be necessary for tissue regeneration to use these functionalized fibrous scaffolds to simulate the extracellular matrix and tissue microenvironment in greater detail. Regardless, coaxial electrospinning adds a dimension to the process that may play a significant role in the development of many tissue engineering products.

3.1. Blood Vessel Scaffolds

The tremendous world-wide clinical need for improved blood vessel substitutes, especially for small-diameter artery prosthetics or grafts, has been the driving force behind a major focus in the field of tissue engineering on developing vascular substitutes. Blood vessel segments with well characterized structure and function are required to repair arterial vessel defects from trauma and disease while minimizing the risk of failure due to thrombus formation and intimal/anastomic hyperplasia. Many recent approaches have incorporated the electrospinning process in vascular prosthetic fabrication, which allows for the modulation of fiber diameter, orientation, and density (Cleary et al.). The fibrous architecture of the prosthetic then influences macroscopic biomechanical properties, contributing to the overall functional performance of these scaffolds (Roh et al., Vaz et al.). It has been theorized that fiber orientation may improve directional migration and anastomotic ingrowth of the native endothelium in animal models (human models to be determined), and the dense fiber meshwork will facilitate cell attachment and spreading (Vaz et al.). Investigations are currently underway regarding the optimization of vascular scaffold architecture and composition to impart overall success post-implantation.

3.1.1. Materials for the Fabrication of Small Diameter Vascular Grafts

Electrospun scaffolds for vascular tissue engineering have been fabricated from natural materials such as collagen types I and III (which typically compose arterial extracellular matrix) (Boland et al., He et al.), elastin (another major component in arterial extracellular matrix) (Boland et al. 2004, McClure et al. 2009), gelatin (Detta et al.), and silk protein (He et al.). Synthetic polymers including poly(lactic acid) (Stitzel et al., T. Wang et al., Su et al.), polycaprolactone (Tillman et al., de Valence et al.), and polydioxanone (Boland et al. 2005, McClure et al.) have also been used, as well as various blends and copolymers. In general, while their mechanical properties are sound, the biocompatibility and results of synthetic electrospun scaffolds have been disappointing, while electrospun natural polymers have shown better biocompatibility. However, scaffolds fabricated from these natural polymers suffer from poor mechanical properties, leaving many to attempt using blends of the two to create scaffolds with the desired properties of both.

From a historical perspective, Stitzel et al. were the first to fabricate a small diameter vascular scaffold via electrospinning of a biodegradable polymer using poly(lactic acid). The vascular scaffolding that was investigated in this study was constructed from collagen fibers (Catgut suture) wound around a stainless steel mandrel and enclosed within electrospun poly(lactic acid) fibers. Subsequently, in 2004, Boland et al. pioneered the electrospinning of micro- and nano-fibrous scaffolds from the natural polymers collagen and elastin for biomimicking vascular tissue engineered constructs. Another example of natural polymers being used in electrospun vascular scaffold fabrication is a study by Soffer et al. in which they fabricated a graft with 5 mm internal diameter from electrospun silk fibroin. This scaffold had a burst strength of 811 ± 77.2 mmHg, approximately half of the burst strength of arterial segments.

The use of electrospun synthetic biodegradable polymers has been extremely popular for vascular tissue engineering applications. In one example, Nottelet et al. fabricated electrospun polycaprolactone small diameter vascular grafts. The results of their study demonstrated that the vascular graft tensile strength was 4 MPa with an elongation at break of 200%. As a frame of reference, both of these values are much higher than the strength and elongation of grafts fabricated from the natural materials. Once these vascular grafts were implanted in the rat abdominal aorta replacement model, it was observed that the grafts remained after the tested 12 week time frame. He et al. fabricated electrospun vascular grafts using elastic, biodegradable poly(L-lactic acid)-co-polycaprolactone and demonstrated a scaffold tensile strength of 7 MPa with an elongation at break of 300%. They also implanted these vascular grafts in the rat abdominal aorta replacement model, and observed that endothelial cells were not present within the lumen of the graft after 7 weeks. Hong et al. fabricated a composite electrospun small vascular graft of poly(etherurethane urea) and poly(2-methacryloyloxyethyl phosphorylcholine-co-methacryloyloxyethyl butylurethane) in an attempt to mimic native arterial mechanical properties. Their mechanical property analysis indicated that the composite scaffolds had a tensile strength of 8 MPa with an elongation at break of 300%. Implantation into the popular rat abdominal aorta replacement model demonstrated the formation of a neo-intima with endothelial cell coverage after 8 weeks along with good anastomic tissue ingrowth from the adjacent arterial segments. Again, it should be noted that the mechanical properties of the electrospun synthetic biodegradable vascular grafts have been shown to be higher than the grafts fabricated from the natural materials.

As demonstrated, the synthetic polymer scaffolds often have the desired arterial-mimicking mechanical properties, but their overall biocompatibility is not sufficient to facilitate vascular regeneration. Consequently, the fabrication of small diameter vascular grafts from blends or composites of both natural and synthetic polymers has garnered serious attention. One primary reason for this attention is that it has been shown that blending a natural polymer (i.e., collagen) with a synthetic results in the segregation of the natural polymer at the electrospun fiber surface. More specifically, only a small fraction (down to approximately 1% in solution prior to electrospinning fibers) is required to introduce functionalization of the fiber surface that mimics that of the pure collagen scaffold response (Chiu et al.). These collagen molecules on the fiber surface exhibit the native integrin binding sites which enhance cell adhesion and migration (Li et al.).

As an example of blended-composition fiber scaffolds, Lee et al. blended polycaprolactone and type I collagen in a ratio of 1:1 in 4.75 mm diameter electrospun vascular grafts (S.J. Lee et al.). The pressurized burst strength of the fabricated grafts was 4,912 ± 155 mmHg, which was significantly greater than that of the control group spun solely of polycaprolactone (914 ± 130 mmHg). Tillman et al. further applied similar polycaprolactone/type I collagen grafts in a rabbit aortoiliac bypass surgery model. The result of this study showed that the electrospun, blended grafts were able to preserve their structural integrity over a 1 month implantation period. In yet another example, Xiang et al. obtained an electrospun small diameter vascular graft composed of two natural materials: recombinant spider silk protein and gelatin with polycaprolactone. This study demonstrated that the seeding of endothelial cells on the luminal surface of the graft resulted in better cellular morphology and cell proliferation rates than those of pure polycaprolactone grafts. This result illustrates the capacity of the natural polymers to enhance cellular proliferation. Du et al. fabricated scaffolds via gradient electrospinning, forming composite fibers of chitosan and polycaprolactone which were impregnated with heparin and endothelial growth factor to reduce the risk of thrombosis and promote the formation of a neo-intima.

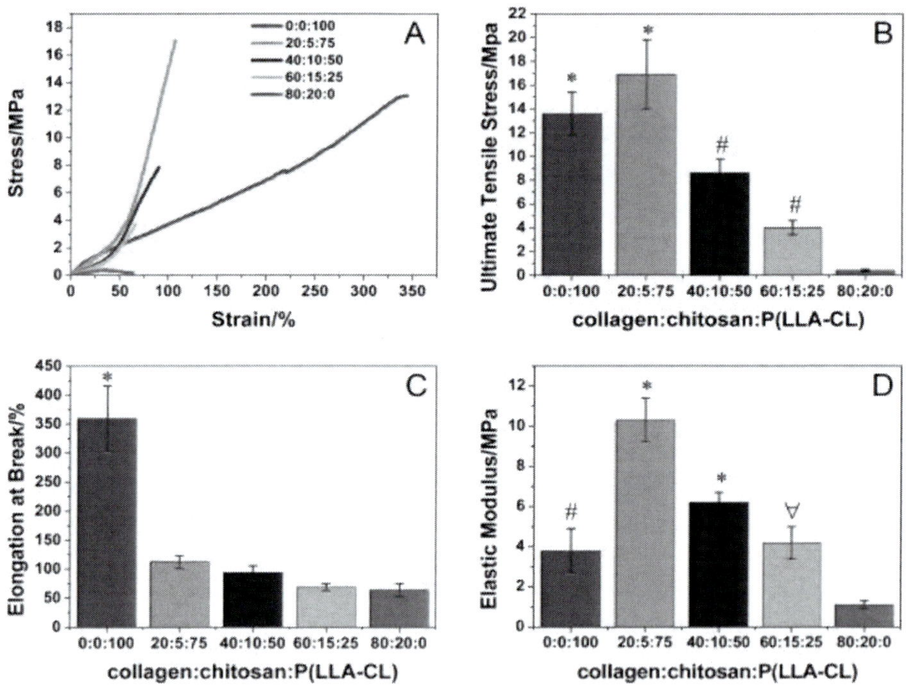

Figure 27. Comparison of mechanical properties of scaffolds with different ratios of collagen:chitosan: poly(L-lactic acid)-co-polycaprolactone. (A) The stress–strain curves of scaffolds. (B) Tensile stress of scaffolds. *indicates a significant difference from 40:10:50, 60:15:25, and 80:20:0 ($p < 0.01$), #indicates a significant different from others ($p < 0.01$) (C) Elongation at break. *indicates a significant difference from other scaffolds ($p < 0.01$). (D) Modulus. *indicates a significant difference from other scaffolds ($p < 0.01$), #indicates a significant different from other scaffold except 60:15:25 ($p < 0.01$), ∀ indicates a significant different from other scaffold except 0:0:100 ($p < 0.01$). (Yin et al. Reprinted with permission from John Wiley and Sons).

The result demonstrated that these structures possessed anticoagulative properties and promoted rapid reendothelialization. Yin et al. fabricated collagen/chitosan/poly(L-lactic acid)-co-polycaprolactone tri-component composite fibrous scaffolds with the addition of a collagen-chitosan component aimed at increasing the overall biocompatibility of vascular grafts (A. Yin et al.). It was also found that by the introduction of a small amount of natural materials (collagen-chitosan), both the ultimate strength (17 MPa) and burst strength (3,000 mmHg) of the collagen/chitosan/poly(L-lactic acid)-co-polycaprolactone composite grafts were higher than the ultimate strength (13 MPa) and burst strength (1,350 mmHg) of pure poly(L-lactic acid)-co-polycaprolactone grafts (Figure 27).

3.1.2. Altering the Microstructure of Small Diameter Vascular Grafts

The blending of natural materials and synthetic materials provides an ideal way for choosing the materials of small diameter vascular graft fabrication. However, the focus in the field has been on efforts to fabricate scaffolds with well-defined, controlled porosity and pore size. The ultimate aim of these efforts is to provide a microenvironment conducive to three-dimensional cell migration and growth into implanted small diameter vascular graft structures in the hope of regenerating a neo-artery. One such effort by McClure et al. in 2012 focused on the design of a porous mandrel to receive the electrospun fibers, which creates a scaffold with a well-defined and patterned vascular graft wall porosity. These vascular grafts had higher porosity in specified regions and a larger pore size due to the introduced air flow (air-impedance electrospinning) being ejected from the porous mandrel. It was shown that the grafts produced had a higher porosity, larger pore sizes in the defined regions, increased water permeability, and cellular infiltration after cell seeding as compared to a traditionally electrospun graft on a solid mandrel (McClure et al. 2012). de Valence et al. in another study fabricated a bi-layer electrospun small diameter vascular scaffold which consisted of a low porosity dense fibrous inner layer and a high porosity fibrous outer layer. Upon implantation of this graft in a rat abdominal aortic replacement model, the dense inner layer played a role in preventing blood leakage and contributed to the formation of a neo-intima, while the porous outer layer provided a microenvironment for the surrounding cells to infiltrate and regenerate a neo-vessel (de Valence et al.).

3.1.3. Anticoagulation of Small Diameter Vascular Grafts

Since the current focus has been on vascular graft structural design and fabrication, researchers rarely consider the issue of incorporating anticoagulants during the electrospinning process. This may be a critical issue to be considered as the work translates to clinical use. Additionally, the majority of the efficacy tests of these fabricated prototypes utilize small animal models (i.e., rat aorta replacement model) (Nottelet et al., Hong et al., W. He et al.). These measures do not meet the real need of small diameter artery regeneration testing and translation

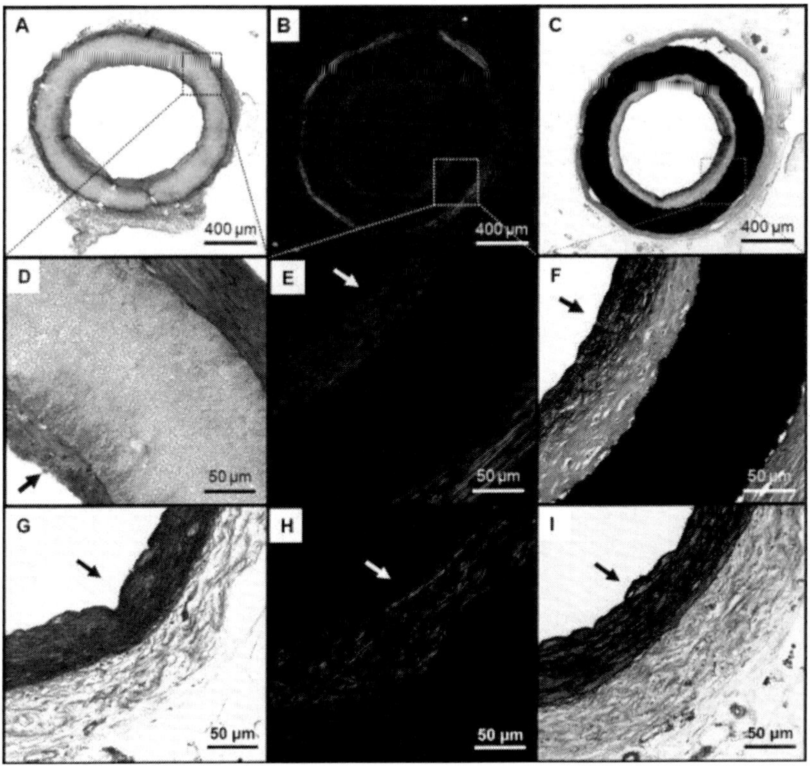

Figure 28. Histological sections of PMA-coated ES-PEUU scaffolds after 12 *in vivo* implantation in rat abdominal aorta. Infiltration of extracellular matrix from both perivascular tissue and lumen is apparent, suggesting scaffold degradation. Samples are stained with (A, D, G) Masson's trichrome staining, (B, E, H) Picrosirius red staining, (C, F, I) Van Gieson staining. G, H, and I are native rat aorta, all other images are implanted scaffolds. (Soletti et al. Reprinted with permission from John Wiley and Sons).

to human use. Thus, a focus on the use of small diameter arteries of larger animals (i.e., canine and porcine) must occur to allow the collection of meaningful data and future clinical translation. In fact, when successful small diameter vascular grafts from small animal models are used in small diameter models in larger animals, such as dog's femoral artery, a high degree of thrombosis will occur without systemic anticoagulation. Armed with knowledge of this shortcoming, many research efforts have focused on improving the capability of anticoagulant incorporation into the fabricated grafts during fabrication or prior to implantation by a variety of methods. As one example, Soletti et al. applied a copolymer (2-methacryloyloxyethyl phosphorylcholine) of ethoxy methyl acryloyl chloride and phosphorylcholine as anticoagulant material on the lumen of electrospun polyether poly(urethane urea) vascular grafts for anticoagulation functionality. The results from this treatment indicated that there was no thrombosis formation on the vascular grafts after implantation in a rat abdominal aorta replacement model for eight weeks and a patency rate of 92% (Figure 28) (Soletti et al.). In contrast, the electrospun polyether poly(urethane urea) control grafts without the anticoagulant coating had an eight week patency rate of 40% with a large degree of thrombus within the lumen. In another study, Hashi et al. evaluated the use of herbal hirudin for anticoagulation by conjugating hirudin via a poly(ethylene glycol) linker to the fibers of an electrospun poly(L-lactic acid) small diameter vascular graft (Hashi et al.). After implanting the poly(L-lactic acid) grafts with and without the hirudin treatment in a rat carotid artery, it was determined that the patency rate of the graft without hirudin treatment was 50%, while that of the hirudin treatment graft was 83% at one month. Thus, this study suggests that this treatment improved the short-term patency.

3.1.4. Implantation of Small Diameter Vascular Grafts in Preclinical Models

To test their safety and efficacy (this aspect often limited in animal model testing), small diameter vascular grafts are evaluated in a variety of preclinical tests using animal models. In one study, Tillman et al. used electrospun polycaprolactone and collagen composite grafts implanted in a rabbit aortoiliac bypass model to evaluate the graft's capacity to support arterial regeneration and withstand the physiological biomechanics. The results from this one month implantation testing demonstrated that the electrospun grafts were able to maintain their structural integrity (i.e., no aneurysm formation), mimicking the native artery. Histological analysis of the explanted vascular graft samples demonstrated that there was minimal cellular infiltration into the graft wall with an absence of an inflammatory infiltrate (Tillman et al.). In the study published by Hashi et al. which evaluated hirudin-treated poly(L-lactic acid) electrospun vascular grafts, they used a rat carotid artery implantation model with 1 mm internal diameter grafts implanted for one and six months. Their results, as noted, showed that they were able to reduce platelet adhesion and aggregation as well as improved graft patency. More importantly, the hirudin-treated grafts displayed improved remodeling and integration along with enhanced endothelialization (complete endothelial cell coverage at one month) and little sign of intimal hyperplasia.

There was new tissue formation observed on the outer surface of the graft, which at one month was relatively loose and highly porous. At six months this tissue became dense with extracellular matrix reorganization and preferential alignment in the circumferential direction. Smooth muscle cell organization (aligned with the circumferential extracellular matrix components) within the neo-artery is critical to its functionality. At six months, the smooth muscle cells in this study were highly organized in the circumferential alignment. In the study by Soletti et al., rat abdominal aorta implantation of methacryloyloxyethyl phosphorylcholine-treated poly(ester urethane) urea vascular grafts demonstrated not only a reduction in platelet adhesion as compared to the non-treated grafts, but also had aligned collagen and elastin similar to native arterial tissue (Figure 29). However, low levels of cellular infiltration were observed, most likely due to low scaffold degradation rate, small pore size, fiber orientation, and scaffold thickness (Soletti et al.). It is thus likely that using this non-thrombogenic coating on a faster degrading scaffold that is more conducive to cellular infiltration would produce a more effective long-term vascular graft (Figure 30). In another study, de Valence et al. implanted electrospun polycaprolactone vascular grafts which consisted of a high-porosity outer layer and a dense inner layer in a rat abdominal aorta replacement model. The results after three weeks of this particular implantation study demonstrated that 32% of the luminal surface was covered with endothelial cells, and this coverage increased to 96% after twelve weeks (de Valence et al.). Their overall conclusion was that a three month period was required for complete luminal coverage of endothelial cells to form a neo-intima. However, intimal hyperplasia development was an issue with this graft design. At three weeks, the luminal coverage of intimal hyperplasia was 8%, and increased to 47.5% after twelve weeks. In addition, this graft wall structure demonstrated cell infiltration and vascularization within the outer layer.

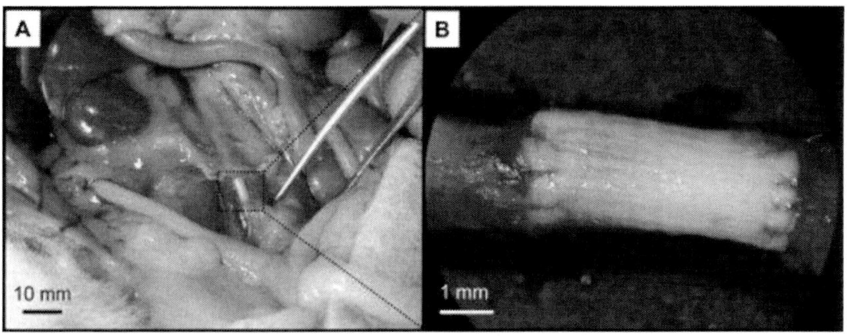

Figure 29. Image of surgically-implanted methacryloyloxyethyl phosphorylcholine-treated poly(ester urethane) urea vascular graft. (A) Scaffold was placed 1 cm downstream of renal arteries, 1cm upstream of the iliac bifurcation. (B) 10-0 Prolene stitches used to anastomose the scaffold to the rat aorta. (Soletti et al. Reprinted with permission from John Wiley and Sons).

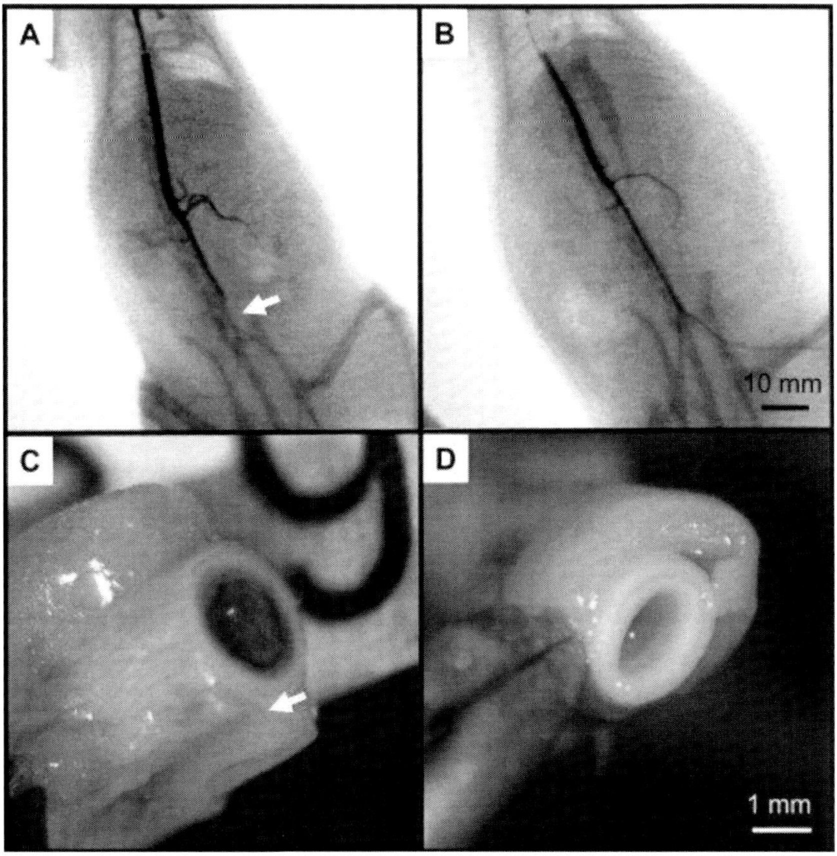

Figure 30. Images taken after 8-week *in vivo* period. (A) Obstructed graft that did not have the anti-thrombogenic coating. (B) Unobstructed graft with anti-thrombogenic coating. (C) Gross pathology image of obstruction in uncoated graft. (D) Gross pathology image of coated graft. (Soletti et al. Reprinted with permission from John Wiley and Sons).

The cell infiltrate was composed of mainly fibroblasts and macrophages, and it is theorized that the vascularization detected was most likely promoted by the angiogenic factors secreted by the resident macrophages. This same group has also performed long-term implant testing (up to eighteen months) of electrospun polycaprolactone grafts in a rat abdominal aorta replacement model (de Valence et al.). After three months, the lumen of the implanted grafts was completely reendothelialized. Intimal hyperplasia was observed simultaneously, and completely covered the entirety of the graft lumens at six months. Similarly, the outer portion of the graft wall was infiltrated by fibroblasts and macrophages, and vascularization occurred simultaneously. Of particular note, the amount of infiltrated cells gradually decreased after six months, and the vascularization also gradually vanished. Additionally, the graft structures were partially degraded and replaced by extracellular matrix produced by the indwelling cells. Therefore, with this structure a complete neo-vessel was not obtained even after eighteen months. Beyond this particular study, there is little research conducted regarding electrospun small diameter vascular grafts implanted in preclinical animal models for longer than six months, and there is little consistency amongst the overall results obtained. Therefore, there remains a very real need for studies of electrospun small diameter vascular grafts implanted for over six months in a reproducible, large animal model to better simulate clinical use in safety and efficacy testing. These studies would allow for concise investigation of the interaction between the vascular graft and surrounding environment, as well as the capacity for vascular tissue regeneration.

3.1.5. Endothelialization of Electrospun Small Diameter Vascular Grafts in vivo

Over the years, there have been a variety of electrospun small diameter vascular grafts implanted in different animal model defects, with varying degrees of endothelium, neo-intima formation at a certain time (de Valence et al., Hashi et al.). Of particular concern regarding this research is that none of the groups conducted studies elucidating the source of the endothelial cells that develop this neo-intima. This neo-intima development will be a critical factor as the electrospun vascular grafts move closer to larger animal models demonstrating potential human efficacy and clinical applications (Zilla et al.). One study by Zheng et al. has provided evidence to the source of endothelial cells lining their implanted electrospun grafts. In this study, they implanted polycaprolactone and polycaprolactone RGD-functionalized grafts into the rabbit carotid artery model. After four weeks, the grafts were explanted and the luminal surface analyzed by scanning electron microscopy to determine the extent of endothelium coverage. Near the anastomosis, both graft types possessed a lumen completely covered by a layer of endothelial cells. In contrast, the mid-graft regions were not completely covered by the endothelial cells, and the endothelialization rates of polycaprolactone grafts were significantly reduced compared to the polycaprolactone RGD-functionalized grafts. The results clearly indicated that the endothelial cells were derived by migration from the anastomotic region and adjacent vessel onto the grafts, and that RGD modification enhanced the migration of the endothelial cells (Zheng et al.). This is a critical aspect of testing in that this degree of anastomotic ingrowth to reendothelialize a vascular graft is unlikely and challenging due to the length of a typical vascular graft (40-60 cm) used clinically. This length is large in comparison to the small animal studies using extremely limited lengths (1-2 cm) in clinical use. Another limitation of these small animal models is that the endothelial cells can migrate gradually from the suture sites to the mid-graft region due to low occurrence of thrombosis. Moving to a large animal model and longer arterial defect replacement models, thrombosis will have a greater potential and this will not allow sufficient time for this anastomotic ingrowth to form a significant neo-intima. Thus, the field may be better served by focusing on transmural capillary ingrowth and blood-borne endothelial cell progenitor cells as potentially clinically feasible methods of deriving a neo-intima on electrospun vascular grafts.

3.2. Bone Scaffolds

Currently, the treatment of large bone defects with medical-grade materials is widely performed with some degree of clinical success. The use of a biomaterial scaffold along with incorporated cells is considered a promising and alternative therapy to the autologous bone graft surgery. The electrospinning process and the fibrous matrices fabricated have thus gained tremendous interest. In this arena, natural biomaterials (e.g., gelatin, silk fibroin, collagen, chitosan), synthetic polymers (e.g., polycaprolactone, poly(L-lactic acid), poly(3-hydroxybutyrate)), and inorganic particles (e.g., bioactive glass, hydroxyapatite, silicate, carbon nanotube, calcium) have been used extensively to fabricate scaffolds mimicking the native bone. Since the extracellular matrix of bone is mainly composed of collagen (organic component) and hydroxyapatite (inorganic component), composite scaffolds for bone tissue engineering are the subject of countless investigations with some of the more recent examples including gelatin/polycaprolactone, polycaprolactone/collagen/hydroxyapatite, poly(lactic-co-glycolic acid)/multi-walled nanotube/ hydroxyapatite, hydroxyapatite/chitosan, hydroxyapatite/poly(L-lactic acid). The presence of bone bioactive inorganic components within biomaterials is generally regarded to favor calcium phosphate mineralization followed by an osteogenic differentiation process. A new strategy for designing electrospun fibers involves providing biological functionality onto the surface of the fibers. Because the cells first recognize the surface of the material, this surface modification will regulate their responses. Modulation of a polymeric surface with materials that are more conducive and active to bone cells, such as a bone mineral-like phase, is just one example of surface tailoring methods targeted for bone regeneration. The incorporation of these inorganic phases can provide nuclei sites for the initiation of mineralization, and the release of ions (such as calcium, phosphate, and silicon) from the inorganic component helps in the precipitation of surface minerals. Moreover, electrospun fibers that are surface-conjugated or incorporated internally with proteins and genes are an elegant way of utilizing these scaffolds in drug delivery to enhance bone regeneration.

3.2.1. Materials for Electrospun Bone Scaffolds

As a natural polymer source, the various types of collagen have long been studied for electrospinning into fibrous scaffolds, in particular for bone scaffolds (Matthews et al. 2003, Jang, Castano and Kim) due to the fact that type I collagen is the main organic, extracellular matrix component of native bone. Thus, it is logical that the electrospinning of type I collagen has drawn a considerable amount of attention for use as a bone regeneration supporting matrix. As one example, Chan et al. has studied mesenchymal stem cell proliferation on electrospun collagen fibers as well as electrospun poly(L-lactic acid)-co-polycaprolactone fibers which were collagen coated. The results of this study demonstrated that the stem cells seeded on the collagen fibers or collagen-coated fibers had enhanced cell spreading with pseudopods attaching to and extending along the fibers with some retention of globular features remaining after 30 minutes of interaction. These data clearly demonstrated that the initial adhesion of these stem cells was dependent on surface texture and the surface chemistry. In another study, Park et al. fabricated electrospun silk fibroin scaffolds with controllable pore size to evaluate their potential application in bone tissue engineering utilizing osteoblast seeding as compared to a commercially available porous poly(lactic acid) scaffold. The results of this study demonstrated that the electrospun silk fibroin supported significantly higher osteoblast proliferation and alkaline phosphatase activity than the control poly(lactic acid) scaffold in their *in vitro* assays. Furthermore, their study results indicated higher expression levels of activated adhesion-related proteins, including focal adhesion kinase, after seeding the silk fibroin scaffolds. These scaffolds were then placed in a rat calvaria model of a critical bone defect in which, after seven weeks, the electrospun silk scaffolds demonstrated a mean of 78% of the original bone volume was regenerated, whereas the control poly(lactic acid) scaffolds resulted in only 49% regeneration of the original bone volume.

Among all synthetic biodegradable polymers, a group of poly (α-hydroxyl acid)s, such as poly(lactic acid), poly(glycolic acid), polycaprolactone and their copolymers, have been the most extensively studied electrospun fibrous scaffolds examined for bone tissue engineering applications. Electrospun polycaprolactone was first suggested as a degradable fibrous matrix for the bone regeneration in a study by Yoshimoto et al. The results from this study demonstrated that neonatal rat mesenchymal stem cells seeded on the electrospun scaffolds migrated into the scaffold and synthesized a collagen extracellular matrix throughout the scaffold. During this regeneration period, it was noted that the cell-polymer constructs maintained the size and shape of the original scaffolds, and no shrinkage was observed. More importantly, after 4 weeks, the cell-polymer constructs had become noticeably harder, indicating that these constructs were potential scaffolding for bone tissue engineering. In another study utilizing electrospun polycaprolactone scaffolds for bone tissue engineering, Shin et al. seeded rat mesenchymal stem cells on the scaffolds and cultured them in a rotary bioreactor containing osteogenic media supplements for four weeks, followed by the implantation of these cultured constructs in rat omenta for an additional four weeks. One key result from this study was that the highly porous scaffolds facilitated cell migration and nutrient and waste transport. These constructs also maintained their original size and shape throughout the study while taking on a rigid and bone-like appearance, suggesting that the unique scaffold topography aided the process of bone development. As an example utilizing electrospun poly(lactic acid) scaffolds, Badami et al. conducted a study using fiber diameters ranging from 0.14 to 2.1 μm to analyze the effect of topography on MC3T3-E1 osteoprogenitor cell proliferation and differentiation. The results of this study showed that electrospun scaffold topographical features do regulate cellular spreading, orientation, and proliferation. More specifically, the actin filaments of the adhered cells appear within both the spin-coated (control film surfaces/planar of equivalent polymer composition) and electrospun substrates, with the difference being that on the electrospun substrate the stress fibers extend perpendicularly across of series of nine fibers and are roughly oriented parallel to one another. The study also noted that in the absence of osteogenic factors, the cell density after seeding was lower on the fibrous scaffold as compared to the film surfaces, while in the presence of osteogenic factors the cell density on fibrous scaffolds was equal to or greater than that on the film surfaces, suggesting the possible influence of osteogenic factors on the osteoblastic responses to the electrospun scaffolding topology.

In another study, Kim et al. electrospun nanocomposite fibers of poly(lactic acid) and hydroxyapatite (HA) to create bone regeneration templates. Using 12-hydroxysteric acid (HAS) as a surfactant, solutions of poly(lactic acid) and HA were spun from a melt to create poly(lactic acid) nanofibers with embedded HA particles, which then were seeded with osteoblasts to determine the cellular response (Figure 31). TEM imaging demonstrated good distribution of the HA powder within the fiber, and cell studies showed higher cell viability and attachment of osteoblasts as compared to poly(lactic acid) fibers without HA incorporated and to culture dish control (Kim et al.).

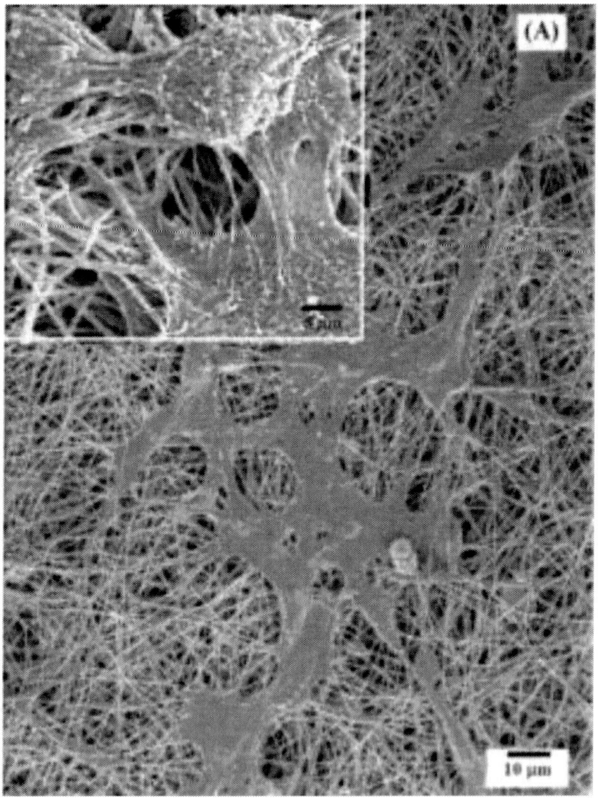

Figure 31. Cellular response of osteoblasts on HA-PLA nanocomposite fiber scaffold after two days post-seeding. (Kim et al. Reprinted with permission from John Wiley and Sons).

To demonstrate the capacity of blending natural and synthetic polymers for improved biocompatibility in bone tissue engineering applications, Song et al. investigated the use of electrospun poly(vinyl alcohol)/collagen/hydroxyapatite fibrous tissue engineering scaffolds. In this study, the hydroxyapatite nanorods were incorporated to reinforce the poly(vinyl alcohol) fibers while collagen was added to enhance cellular interaction. The results showed that these additives created organic-inorganic fibers capable of degradation while reinforcing the hydrolytic resistance (provided longer lasting stability) and mechanical properties of the electrospun fibrous scaffolds. These two encapsulated components also enhanced the adhesion and proliferation of murine bone cells (MC3T3) under *in vitro* culture conditions. Zhang et al. conducted a study using electrospun chitosan/hydroxyapatite fibrous scaffolds using a two-step method. The results of this study indicated that the hydroxyapatite particles incorporated within the scaffolds, as compared to chitosan control scaffolds, resulted in significantly increased bone formation capacity (increased cell proliferation, mineral deposition, and morphology observation). Zhang and Chen fabricated electrospun poly(lactic-co-glycolic acid)/multiwalled carbon nanotube/ hydroxyapatite composite scaffolds for bone tissue engineering. The study indicated excellent biocompatibility and that the attachment and proliferation of rat bone marrow-derived mesenchymal stem cells on the composite scaffolds was significantly increased over poly(lactic-co-glycolic acid) scaffolds.

3.2.2. Surface Modification of Bone Scaffolds

Since electrospun fibrous scaffolds have an extremely large surface-area-to-volume ratio, the surface-related properties are critically important in determining the cell interaction. It is vital, then, to explore the tailoring of the electrospun fiber surface properties by post-processing techniques which introduce surface characteristics. These surface characteristics drive the desired biological responses and improve the overall potential of electrospun fibers as a cell matrix analogue scaffolding. The main methods for surface modification include plasma treatment, surface functionalization by coating with bioactive materials, and direct encapsulation.

Oxygen plasma treatment was first introduced to tailor the nano-scale surface properties (roughness) of electrospun polymeric fiber structures in order to create an improved bone tissue engineering scaffolding (Nandakumar et al.). The results showed that both fiber nano-roughness and the surface oxygen content can be tuned by controlling the duration of plasma treatment. While no significant effect of plasma treatment was seen on the attachment and proliferation of clinically relevant human mesenchymal stem cells, a positive effect was derived on the plasma treated surfaces in terms of the stem cell expression of osteogenic genes osteonectin and bone sialoprotein after 7 days of culture. Additionally, the stem cell alkaline phosphatase was most pronounced after 21 day culture on the plasma treated scaffolds as compared to untreated scaffolds. In contrast, a study by Martins et al. cultured fibroblasts, chondrocytes, and osteoblasts on polycaprolactone scaffolds subject to either oxygen or argon plasma treatment, and showed that oxygen treatment at 30W for 10 minutes and argon treatment at 20W for 5 minutes created optimal substrates for cell adhesion and proliferation. While they hypothesize that this effect in the oxygen-treated fibers is due to the increased hydrophilicity of the fibers (Figure 32), the argon-treated fibers showed a similar effect despite their more hydrophobic character, possibly due to increased fiber roughness (Figure 33) (Martins et al.). Thus, while plasma treatment is a useful tool in promoting cell attachment and proliferation, it is not the only factor in this regard.

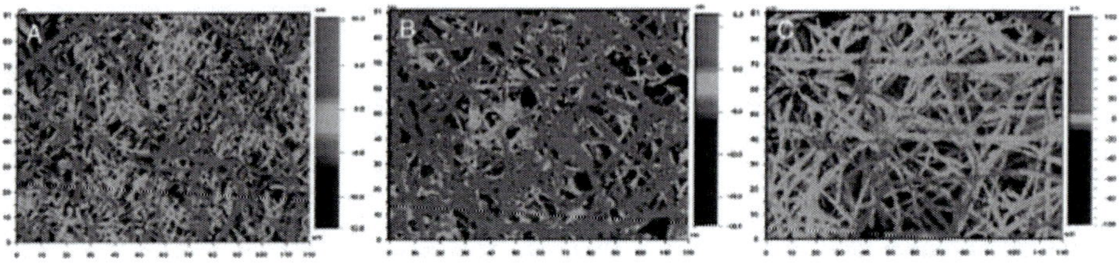

Figure 32. Images of optical profilometry of (A) untreated polycaprolactone nanofibers, (B) oxygen-treated for 10 minutes at 30W, and (C) Ar-treated for 5 minutes at 20W (Martins et al. Reprinted with permission from John Wiley and Sons).

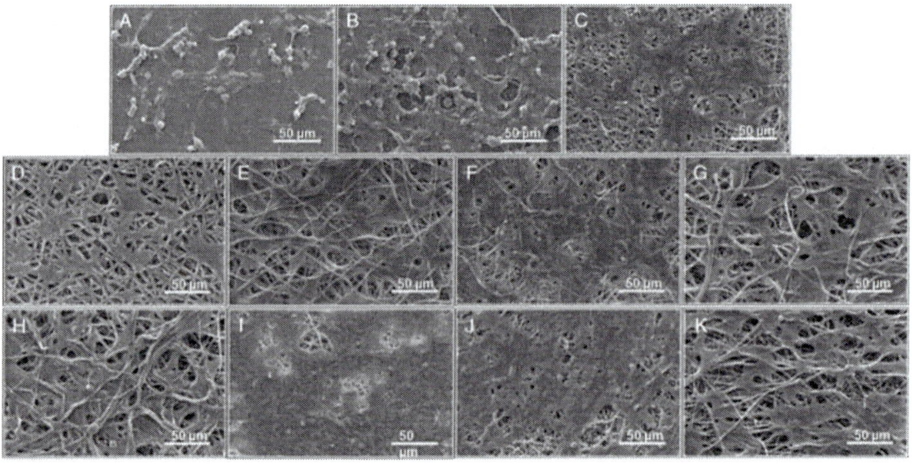

Figure 33. Scanning electron microscopy images of osteoblasts grown on (A) untreated tissue culture polystyrene, (B) solvent-cast polycaprolactone, (C) untreated polycaprolactone nanofibers, (D) polycaprolactone nanofibers treated with oxygen at 30W for 10 minutes, (E) polycaprolactone nanofibers oxygen treated at 30W for 5 minutes, (F) polycaprolactone nanofibers oxygen treated polycaprolactone at 20W for 10 minutes, (G) polycaprolactone nanofibers oxygen treated at 20W for 5 minutes, (H) polycaprolactone nanofibers argon treated at 30W for 10 minutes, (I) polycaprolactone nanofibers argon treated at 30W for 5 minutes, (K) polycaprolactone nanofibers argon treated at 20W for 5 minutes (Martins et al. Reprinted with permission from John Wiley and Sons).

Yang et al. studied electrospun polycaprolactone fibrous scaffolds containing bone-like calcium phosphate coatings which maintained a fibrous, porous architecture. After fabricating the scaffolds, they were immersed and incubated in a simulated body fluid for up to seven days. After this time period, the fiber coating formed on the polycaprolactone transformed into pure calcium deficient type B carbonate apatite with a crystallinity similar to biological apatite. In yet another example of fiber surface modification, Li et al. coated polycaprolactone fibers with gelatin and calcium phosphate to fabricate a bone biomimetic scaffold. The surfaces of the fibers were coated with gelatin through a layer-by-layer self-assembly, followed by functionalization with a uniform coating of bone-like calcium phosphate by mineralization in simulated body fluid for 2 hours. This study showed that preosteoblastic MC3T3-E1 cells attached, spread, and proliferated well with a flattened morphology on the mineralized scaffolds. The proliferation rate of the cells on the mineralized scaffolds was significantly higher (1.9-fold) than that on the untreated fibrous scaffolds after 7 days in culture.

In terms of encapsulating functionalizing factors, one group studied the encapsulation of bone morphogenetic protein 2 (BMP-2) and nano-crystal hydroxyapatite in electrospun silk scaffolds to promote bone regeneration (C. Li et al.). The scaffolds with the co-processed BMP-2 and hydroxyapatite supported an increased calcium deposition and enhanced transcript levels of bone-specific markers when seeded with human bone marrow-derived mesenchymal stem cells for 31 days as compared to the controls (scaffolds without BMP or hydroxyapatite), indicating that these electrospun silk scaffolds were capable of maintaining BMP-2 activity and efficiently delivering BMP-2 to enhance bone regeneration potential. If using BMP and organic solvents in the formation of electrospun scaffolds for bone tissue engineering, care must be taken to ensure the activity of the encapsulated BMP is maintained (Madurantakam et al.). In another study, Nie et al. fabricated BMP-2 deoxyribonucleic acid (DNA) plasmid loaded poly(lactic-co-glycolic acid)/hydroxyapatite composite scaffolds via electrospinning with the DNA incorporated into the scaffolds in three different processing methods (1. Naked DNA, 2. Encapsulation/entrapment of DNA/chitosan nanoparticles into fabricated scaffolds, and 3. Encapsulation of DNA/chitosan nanoparticles into scaffold fibers during electrospinning). The results of cell culture experiments with human bone marrow-derived mesenchymal stem cells demonstrated that the fiber-encapsulated plasmid/hydroxyapatite scaffolds had higher cell attachment and viability as well as more desirable transfection efficiency than the other loading configurations. This study indicates that this scaffolding modification method resulted in a composite scaffold with significant promise in bone regeneration applications.

3.2.3. Large Pore Scaffolds

Bone tissue has a three-dimensional (3D), mineralized structure that must be mimicked during attempts to regenerate form and function. In addition, bone is continuously degraded and regenerated by complex interactions between osteoclasts (which break down bone) and osteoblasts (which build bone). The typical small pore sizes in electrospun scaffolds are most often a hindrance to cell infiltration *in vitro* and tissue-ingrowth (cell infiltration and angiogenesis) into the scaffold *in vivo*, limiting their clinical potential. Thus, when designing a bone tissue engineering scaffold the migration of cells into a biodegradable scaffold is a crucial aspect for the overall success and healing of the bone defect. Additionally, the small pore sizes also constrain the vascularization of the scaffolds, which restricts nutrient delivery and waste removal and limits the degree of tissue-ingrowth and regeneration that can be supported (Karageorgiou and Kaplan). This challenge has led to extensive efforts to enlarge the pore size (inter-fiber distance) of electrospun scaffolds.

One example is a study conducted by Park et al. in which they developed a pore-size controllable three-dimensional electrospun silk fibroin scaffold via a salt-leaching method. To enhance inter-fiber spacing during processing, sodium chloride crystal particles of various sizes were co-deposited on the collection system amongst the collecting fibers. After salt leaching by water immersion, the fabricated scaffolds had pore sizes ranging from 0-500 μm, which significantly increased the porosity and the amount of water absorbed by silk scaffolds as compared to traditional electrospun scaffolds. These enhanced porosity scaffolds were then implanted in a critical-size bone defect in a rat calvaria model and then evaluated by microcomputed tomography analysis. After seven weeks, the results indicated that 78% of the original bone volume was regenerated in the large-pore silk scaffold implantation group compared to only 49% for a commercially available porous poly(lactic acid) scaffold. In another study, Phipps et al. utilized sacrificial fibers of polyethylene oxide in a similar fashion to salt crystal leaching to enhance scaffold porosity. More specifically, they electrospun polycaprolactone/collagen/hydroxyapatite scaffolds with the inclusion of sacrificial fibers to effectively increase pore size. The results showed that the enhanced and effective porosity promoted increased human mesenchymal stem cells infiltration as well as greater infiltration of endogenous cells from calvarial organ cultures. In summary, all of these methods of creating electrospun bone scaffolding have shown the critical dependence of introducing the capacity for three-dimensional cell infiltration to allow for functional bone regeneration.

3.3. Nerve Regeneration

Peripheral nerve injuries are very common in clinical practice, especially in trauma cases. In these cases, union of severed nerve ends remains a challenge and often results in incomplete or even loss of functional recovery in many patients, subjecting them to lifelong disability. These disabling injuries come at a huge economic demand/loss and physiological burden to the patients and their families. For these reasons, developing a method or system to promote nerve regeneration and restore the distal function has been an area of tremendous focus and efforts. In the case of peripheral nerve injuries such as a severed nerve, the nerve ends (or stumps) are generally repaired (if possible) by direct end-to-end (anastomosis) suture. If the close approximation of the two nerve ends is not feasible for such a direct repair, the clinical "gold standard" for repair remains an autologous nerve graft. However, the use of autologous nerve graft has limitations and causes donor site injury and morbidity. The limitations include a low rate of regeneration, limited availability of nerves for use and lack of coaptation between the injured nerve and the nerve graft most often due to a size mismatch (Cao, Liu and Chew). Therefore, the efforts in developing an autologous nerve graft substitute have garnered significant attention.

When designing a nerve regeneration guidance system, the neurite outgrowth (axonal extensions and growth cone) and stump cell migration are indispensable processes of major attention. Regardless of the design, guiding the axonal extension towards designated targets to reform synaptic connections is critical in the restoration of nerve function. In the nerve injury site, the glial cells which migrate to the site will secrete neurotrophic factors and support neural regeneration. Hence, the ability to enhance neurite outgrowth, direct axon extension, and facilitate cell migration while providing a microenvironment conducive to regeneration are essential factors to be included in the design of scaffolds for nerve repair. Thus, when designing a nerve regeneration guidance system, one must maintain a well-defined, continuous pathway and space (to allow nutrient diffusion) for regeneration, promote infiltrating cells to secrete factors that induce axonal elongation with minimized scar tissue formation, relieve the overall tension associated with direct end-to-end suturing that alter microenvironment biomechanics, and minimize tissue ingrowth from the surrounding tissue (Prabhakaran, Venugopal and Chyan et al.). The use of electrospinning in fabricating substrates for peripheral nerve repair/regeneration has been widely investigated (F. Yang et al., Nisbet et al.). The most basic of these studies have demonstrated in models of comparisons between substrates made of polymer films versus electrospun fibers that both *in vitro* and *in vivo* results demonstrate clear efficacy of electrospun fibers in enhancing nerve regeneration.

3.3.1. Materials Applied in Nerve Tissue Engineering

As with any tissue engineering scaffold, the material selection plays an essential role in the success of nerve regeneration using electrospun scaffolding. Additionally, an appropriate material degradation rate and structural mechanical properties will favor a minimized inflammatory response and prevent nerve compression situations, whilst providing the required support and guidance for axon extension and regeneration (Prabhakaran, Venugopa and Chyan et al.). Examples of polymers used in electrospun nerve tissue engineering scaffold fabrication include poly(glycolic acid), poly(L-lactic acid), polycaprolactone, polydioxanone, polyhydroxybutyrate, and various co-polymers, with the ability to support cell proliferation and viability differing due to their distinct properties.

In terms of the specific capacity of electrospun polymers for use in neural tissue regeneration, a study conducted by Sangsanoh et al. reported the *in vitro* responses of Schwann cell line (RT4-D6P2T) on various types of electrospun fibrous scaffolds (polyhydroxybutyrate, poly(3-hydroxybutyrate-co-3-hydroxyvalerate), polycaprolactone, poly(L-lactic acid), and chitosan) in comparison to solution-cast films. The overall results of this study demonstrated that upon cell seeding and twenty four hours of culture, there were variations in the cell viability within the materials and structures. The polycaprolactone (film and electrospun fibers) results were equivalent to tissue culture plastic controls with all others responsible for decreasing viability until a minimum on the polyhydroxybutyrate fibrous scaffold at the twenty four hour time period. However, after three days of culture, the rankings in terms of cell viability had slight rearrangements with the tissue culture plastic and films having the maximum viability and the fibrous poly(3-hydroxybutyrate-co-3-hydroxyvalerate) scaffold having the minimal viability of the materials tested. Additional insight gathered from this study was that after eight hours in culture, the cells on the polymeric films and the electrospun polycaprolactone appeared in their characteristic spindle shape. Those on the other electrospun scaffolds, except chitosan, also had similar appearance, but in these cases the cells were seen penetrating into the scaffolds. It was not until day three of culture that the cells on the surface of electrospun chitosan appeared in their characteristic spindle shape for the first time. It should be noted that in this study there were varying average fiber diameters of the different electrospun materials, from 130 nm to 3.7 μm. Thus, more in-depth studies are required in order to understand the mechanisms of how material property may affect neural cell behaviors.

In the field of tissue engineering, the use of natural materials/polymers (i.e., collagen) has shown increased biocompatibility with similar properties (mechanical strength, physical properties and biomolecular recognition sites) to the tissues for which they are being used to regenerate (Willerth and Sakiyama-Elbert). Since these materials are obtained from natural sources, they must be processed with care and purified to ensure that no foreign body response occurs upon implantation. In terms of neural regeneration scaffolds, electrospun type I collagen has been successfully fabricated by using acetic acid as the solvent and incorporating a unique cross-linking agent, rose bengal (T. Liu et al.). In their study, the results demonstrated that the scaffolds provided a favorable microenvironment in support of neural stem cell line C17.2 over a seven day period. This evaluation was based on the cellular capacity to proliferate and form a confluent layer as well as differentiate, as demonstrated by numerous cellular projections which are indicative of neurite outgrowth. These results indicated the potential application of these scaffolds in nerve tissue engineering.

Wang et al. developed electrospun chitosan micro/nanofiber tubes (nerve guides) for use in testing in the treatment of a ten millimeter rat sciatic nerve defect model. After implanation of these scaffolds, it was observed that sensory function had begun to recover in as early as eight weeks using the electrospun nerve guides, although functional recovery of motor activity was delayed. The histological examination of these nerve guides was consistent with the activity detected with regenerating axons seen elongated through the lumen of the electrospun chitosan tubes. In summary, the nerve guides generated and tested had sufficient mechanical strength for use in preserving tube space while providing a scaffold for cell migration and attachment which facilitates enhanced nerve regeneration. In spite of the regeneration obtained by these electrospun scaffolds, the overall weak mechanical properties of natural materials limited their use as nerve guides. In order to maintain structural integrity in many implant sites, sufficient strength will be required over a defined period of time to allow regeneration. Another concern in this application is the higher cost, possible immunogenicity, and batch-to-batch processing and fabrication variability.

In the field, the main approaches to overcome the limitations of natural materials are the blending with synthetic polymers or synthetic polymer fiber surface modification. In one study, Prabhakaran et al. fabricated electrospun polycaprolactone, chitosan, and polycaprolactone/chitosan blended fibrous scaffolds with average fiber diameters of 630, 450, and 190 nm, respectively. The mechanical evaluation of the scaffolds showed that the blending of the two polymers increased the Young's modulus and strain at break over those of just chitosan. The surface analysis of the fibers indicated that the presence of chitosan on the blended fiber surfaces generated greater hydrophilicity. This surface alteration led to enhanced rat Schwann cell attachment and proliferation (48% increase in proliferation) on polycaprolactone/chitosan scaffolds as compared to just polycaprolactone over an eight day culture period, all while maintaining the characteristic cellular morphology along the fibers. In another study, electrospun gelatin/polycaprolactone composite scaffolds (random and aligned) were fabricated with weight ratios of 50:50 and 70:30 for evaluation in nerve regeneration (Ghasemi-Mobarakeh et al. 2008). The results demonstrated that the overall properties of the electrospun scaffolds were strongly influenced by the increasing concentration of gelatin in the composite fibrous structures. More specifically, it was determined that the electrospun 70:30 polycaprolactone/gelatin scaffolds exhibited the most balanced properties, meeting the required specifications for nerve tissue. Thus, these scaffolds were tested in *in vitro* cell culture assays with C17.2 neural stem cells. The results of these cell culture assays indicated that the 70:30 polycaprolactone/gelatin scaffolds enhanced differentiation and proliferation compared to polycaprolactone, representing a positive guidance for potential neural regeneration. Schnell et al. electrospun polycaprolactone and collagen/polycaprolactone aligned scaffolds for testing in cell culture assays. The results of this study showed that both scaffold types were favorable substrates for cell attachment, Schwann cell migration, and axonal regeneration with the presence of 25% collagen significantly enhancing these biological effects. Wang et al. combined poly(lactic-co-glycolic acid), silk fibroin, and collagen to fabricate electrospun composite scaffolds with the resulting mechanical properties strongly influenced by the concentration of collagen and silk fibroin incorporated. More importantly, it was observed that Schwann cells can adhere, migrate, and extend neurites (bipolar morphology) on these scaffolds. In a surface modification study, electrospun polycaprolactone was modified by plasma to alter the surface properties by the introduction of desired functionalities on the surface. This modification improved the hydrophilic properties and permeability (Prabhakaran, Venugopal and Chyan et al.). Schwann cells seeded on these scaffolds showed a comparatively higher proliferation (17% increase) on the plasma-treated polycaprolactone scaffolds as compared to polycaprolactone/collagen scaffolds, while also retaining their normal phenotypic morphology (bipolar elongation along fibers) after an eight day culture.

3.2.2. Effect of Nanofiber Orientation on Nerve Regeneration

While micro- and nano-patterned substrates have been extensively studied to understand the effects of contact guidance on neural cell behavior and function, aligned-fiber electrospun scaffolds have been gaining significance for their potential in providing topographical cues for nerve regeneration (Cao, Liu and Chew). As a key finding, Kim et al. demonstrated that the topographical cues of electrospun scaffolds can influence endogenous peripheral nerve repair mechanisms even in the absence of exogenous growth factors. In their study, the aligned, but not randomly oriented, electrospun fibrous scaffolds successfully promoted regeneration of axons across a 17 mm rat tibial nerve resection model. These axons reinnervated muscles and reformed new neuromuscular junctions. Additionally, the quantitative comparison of dorsal root ganglion neuron outgrowth on the aligned fibers demonstrated that the majority of neurite outgrowth and Schwann cell migration extended unidirectionally, parallel to the aligned fibers. As noted, the aligned fibers facilitated the nerve regeneration of transected tibial nerves across 17 mm nerve gaps.

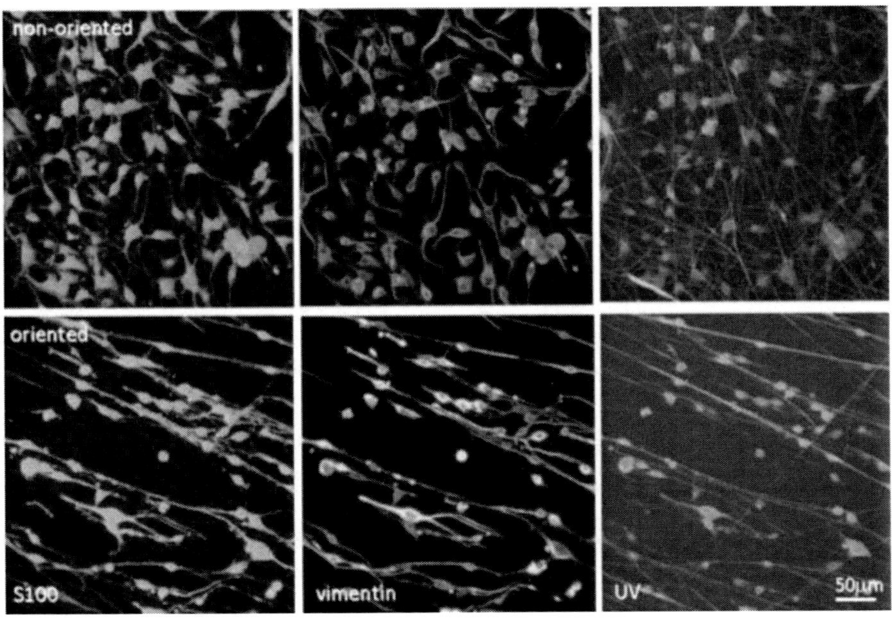

Figure 34. Schwann cells cultured on non-oriented (upper) and oriented (lower) chitosan nanofiber mesh sheets (Wang et al. Reprinted with permission from John Wiley and Sons).

The host-derived Schwann cells infiltrated into the constructs from both proximal and distal nerve stumps. The transected axons entered into the proximal end of the construct and extended through the entire length of the scaffold along the stacked aligned fibers and moved into the distal nerve stump. In a similar study, Wang et al. cultured Schwann cells on electrospun chitosan aligned fiber tubes, and then implanted these constructs as bridge grafts into rat sciantic nerve defects. These tubes were created in three groups: tubes with non-oriented fibers, tubes with fibers oriented along a longitudinal axis, and bilayer tubes with an inner layer of oriented nanofibers and an outer layer of randomized nanofibers. Cell alignment along the fiber axis was seen on the oriented-fiber constructs *in vitro* (Figure 34), and functional electorphysiological recovery was exhibited on the same level of isograft controls. In addition, myelinated axon sprouting was seen on the oriented and bilayer constructs, comparable to the isograft control (Figure 35) (Wang et al.).

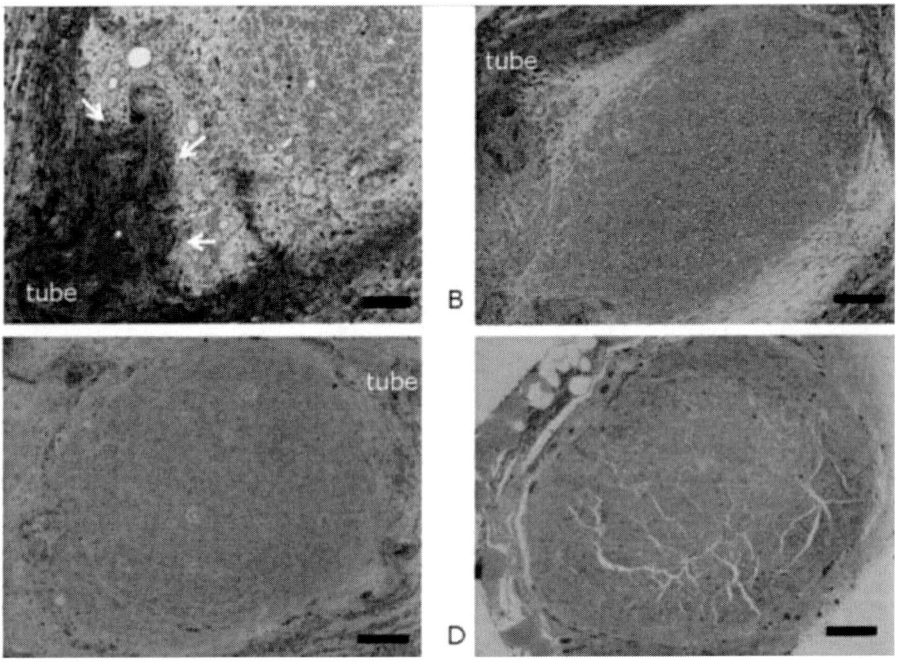

Figure 35. Histological images of grafted chitosan electrospun tubes and isograft after 30 weeks implantation. (A) nonoriented fiber tube, (B) oriented fiber tube, (C) bilayered tube, (D) isograft. The arrows denote areas of separation and collapse of the inner layers of the nonoriented tube. In contrast, the oriented and bilayered groups had good preservation of their inner tube space, creating isograft-like morphology (Wang et al. Reprinted with permission from John Wiley and Sons).

In a study which analyzed the differentiation of embryonic stem cells seeded on randomly and uniaxially aligned electrospun polycaprolactone scaffolds, Xie et al. examined the potential of these scaffolds to enhance the differentiation into neural lineages and promote and direct neurite outgrowth. The overall premise of this study was that the combination of an electrospun scaffold and embryonic stem cells could lead to an enhanced nerve regeneration/repair. The results of this study demonstrated that the uniaxially aligned fibers enhanced the differentiation of the embryonic stem cells into neural lineages but also controlled the direction of neurite outgrowth.

A study by Yang et al. supported these results by showing that scaffolds made of aligned electrospun fibers were better suited for culturing nerve stem cells *in vitro* than scaffolds of random fibers. They cultured neural stem cells on electrospun poly(L-lactic acid) scaffolds of aligned nanofibers (ANF), aligned microfibers (AMF), and random nanofibers (RNF), and demonstrated neural stem cell elongation and neurite outgrowth in the direction of the fibers on the AMF scaffolds. Furthermore, they showed that fiber diameter had no effect on neural stem cell orientation, but that neural stem cells cultured on nanofibers differentiated faster than those on microfibers. Thus, this study indicates that ANF poly(L-lactic acid) scaffolds are optimal cell carriers for neural tissue engineering.

In a similar study, J. Wang et al. grew floating neurospheres of human embryonic stem cell-derived neural precursors (NPs) and then cultured them on RNF and ANF scaffolds of electrospun silk fibroin. Immunostained images showed that culture on ANF induced more neurite extension in the NP neurospheres than the RNF culture, and these extensions were along the fiber axis of alignment (Figure 36). Cells cultured on ANF also migrated away from the neurospheres in the direction of fiber alignment, while the RNF cells stayed closely grouped to their neurosphere. By the second day of culture, a bipolar morphology was observed, and after seven days secondary and tertiary branches were formed. Neurite growth along the fiber axis was longer on the 400 nm diameter fibers than the 800 nm diameter fibers, but aligned to the fiber axis only after an initial period of growth. Fiber alignment was shown to increase the expression of β-III-tubulin on both fiber sizes, but the cells cultured on 400 nm diameter fibers had 6% higher β-III-tubulin than those cultured on 800 nm diameter fibers (Figure 37). Total number of neurites was also highest on 400 nm diameter aligned fibers, and then were found on (in descending order): 400 nm diameter randomized fibers, 800 nm diameter random fibers, and 800 nm diameter aligned fibers. It was thus concluded that the 400 nm diameter aligned fiber scaffold was more suitable for neuronal differentiation than the alternatives examined in this study (J. Wang et al.).

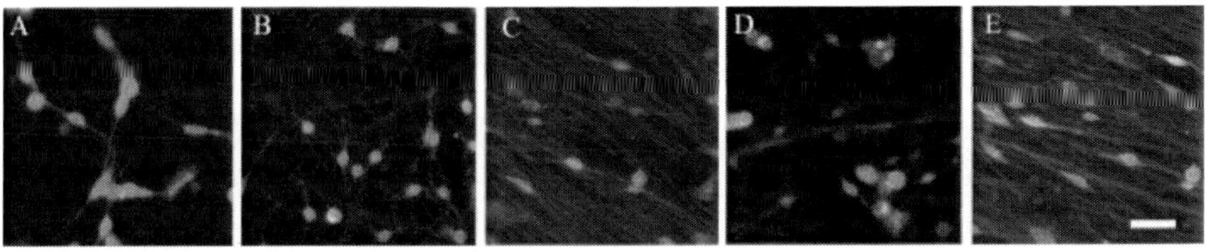

Figure 36. Live/dead-stained dissociated human embryonic stem cell-derived neurospheres cultured on (A) Poly-D-lysine/laminin substrate, (B) random 400 nm diameter nanofibers, (C) aligned 400 nm diameter nanofibers, (D) random 800 nm diameter fibers, (E) aligned 800 nm diameter nanofibers. All fibers spun from silk fibroin. (J. Wang. et al. Reprinted with permission from John Wiley and Sons).

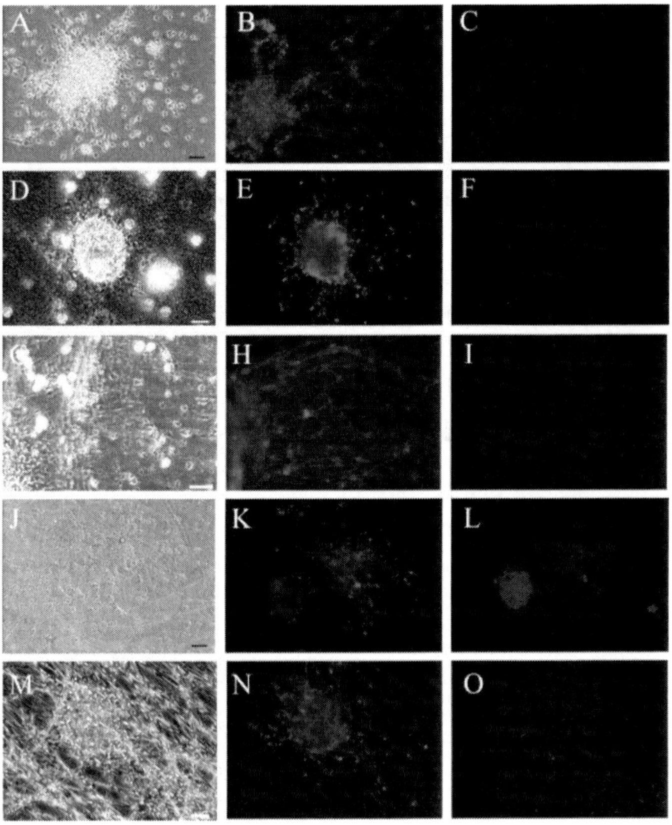

Figure 37. Neurospheres after 7 days of culture, immunostained for β-III-tubulin. Cells are cultured on (A-C) Poly-D-Lysine/laminin, (D-F) randomly oriented 400 nm diameter fibers, (G-I) aligned 400 nm diameter fibers, (J-L) randomly oriented 800 nm diameter fibers, and (M-O) aligned 800 nm fibers. (J. Wang. et al. Reprinted with permission from John Wiley and Sons).

3.3.3. Growth Factors and Conductive Nanofibers

Nerve growth factor has been widely employed in the regeneration of neural tissue because it has shown promise in maintaining neuronal differentiation. In one study, nerve growth factor was loaded into a fibrin matrix for controlled release of the growth factor. When this matrix was applied to peripheral nerve defects, the axonal growth of the neural cells was significantly enhanced in a dose-dependent manner (A. C. Lee et al. 2005). In another study, nerve growth factor was physically immobilized on the surface of cover slips to enhance the proliferation of PC12 cells (Bhang et al.). More specifically, nerve growth factor was encapsulated into or on electrospun fibers by conjugating (Cho et al.) and blending electrospinning and coaxial electrospinning (J. J. Liu et al.). In the study by Cho et al., nerve graft factor was chemically immobilized on electrospun scaffolds which effectively differentiated rat mesenchymal stem cells into neuronal-like cells compared to surface adsorbed NGF. The rat mesenchymal stem cells cultured on nerve growth factor-conjugated electrospun scaffolds with topographical cues demonstrated an elongated morphology and aligned neuron cell-specific cytoskeletons. In another example of nerve growth factor delivery, Valmikinathan et al. fabricated electrospun scaffolds from a blend of polycaprolactone and bovine serum albumin to form fibers for the controlled release of nerve growth factor. The purpose of the albumin was to aid in overcoming the most common disadvantages associated with hydrophobic polymers. These disadvantages include reduced drug loading efficiency, long degradation periods, and an initial burst release. Using an enzyme-linked immunosorbent assay, they were able to demonstrate that the blended fibers had an efficient release of nerve growth factor over a 28 day period *in vitro*. By introducing the released nerve growth factor into the cultured PC12 cells, they were able to demonstrate that the released factor retained its bioactivity for stimulating neurite outgrowth over the 28 day period. Liu et al. developed a composite nerve conduit comprised of poly(L-lactic acid)-co-polycaprolactone and nerve growth factor. This was developed from core-shell structured electrospun fibers, which were fabricated by coaxial electrospinning of poly(L-lactic acid)-co-polycaprolactone as the shell and bovine serum albumin/nerve growth factor within the core. The results from their analysis showed that the controlled release of encapsulated nerve growth factor was able to promote peripheral nerve regeneration and demonstrated that the composite nerve tube was not statistically different in terms of regenerative capacity from the positive autograft control, which has been the clinical "gold standard".

Electrically conductive polymers have also attracted much interest since they simultaneously display the physical and chemical properties of organic polymers and the electrical properties of metals (Chronakis, Grapenson and Jakob). The use of electro-active materials would allow for the local conveyance of electrical stimulus, further providing a physical template for neural cell growth and tissue repair (Rivers, Hudson and Schmidt). In 2009, Ghasemi-Mobarakeh et al. fabricated a conductive electrospun polycaprolactone/gelatin scaffold by doping the fibers with polyaniline and investigated the effect of electrical stimulation of nerve stem cells seeded on these scaffolds.

The electrical stimulation of neural stem cells through the conductive fibrous scaffolds enhanced cell proliferation and neurite outgrowth more significantly than the non-electrically stimulated scaffolds (Figure 38). Xie et al. used a combination of electrospinning and aqueous polymerization to create conductive core-shell nanofibers to allow for the stimulation of dorsal root ganglia cells. Templates were electrospun from polycaprolactone or polylactic acid, and then pyrrole was polymerized on the fiber surface to provide topographic cues and electrical stimulation through the fibers. These templates were created with both random and aligned fiber structures. Dorsal root ganglia cells were cultured on the templates, and it was observed that neurite alignment was much stronger on the aligned-fiber scaffolds. A constant voltage of 10V was applied for 4 h per day across the templates for 6 days of culture (Figure 38). At the end of this period a significant increase in maximum neurite length was shown in the electrical stimulation group as compared to the non-electrically-stimulated control, for both aligned and randomized fiber types. However, this effect was greater on the random nanofiber cell culture as opposed to the aligned nanofiber culture, which was theorized to be due to a finite limit of neurite growth for a fixed period of time. As the neurites cultured on aligned fibers grew longer than those cultured on randomized fibers, they would be closer to this finite limit and thus have less room for enhancement by the electrical stimulation. Nevertheless, the improvement in neurite length with electrical stimulation was significant on both aligned and randomized fibers, indicating the usefulness of this technique (Xie et al.).

Shao et al. fabricated randomly oriented and aligned electrically conductive electrospun fibers of poly-D,L-lactide in which multiwalled carbon nanotubes were embedded. Without electrical stimulation, cell proliferation was slightly increased on the aligned poly-D,L-lactide/multiwalled carbon nanotubes fibrous scaffolds than those on random oriented scaffolds, and osteoblasts grew along the axis of aligned nanofibers. On randomly oriented scaffolds, the inclusion of electrical stimulation significantly promoted elongation of obsteoblasts, while on the aligned scaffolds the stimulation only slightly enhanced elongation along the fiber orientation. In addition, electrospun poly(L-lactic acid)/polyaniline scaffolds were found to possess the nano-scale features of the native extracellular matrix (Prabhakaran et al. 2008). The electrical stimulation of nerve stem cells on these scaffolds showed higher neurite extensions, which facilitate overall nerve regeneration. An image of a carbon nanotube incorporated into an electrospun fiber is shown in Figure 39.

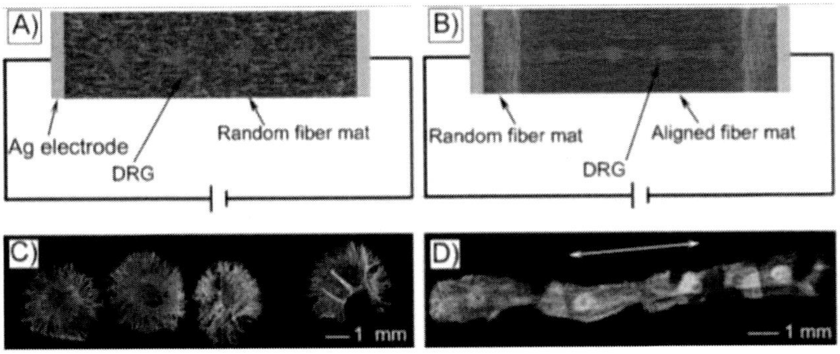

Figure 38. (A and B) Schematic of dorsal root ganglion neurite growth on random and aligned polycaprolactone – polypyrrole core-shell nanofibers under electrical stimulation. (C) Close-up of dorsal root ganglion neurite outgrowth on randomly oriented fibers under electrical stimulation. (D) Close up of dorsal root ganglion neurite outgrowth on aligned fibers under electrical stimulation. The arrow shows the direction of fiber and neurite alignment (Xie et al. Reprinted with permission from John Wiley and Sons).

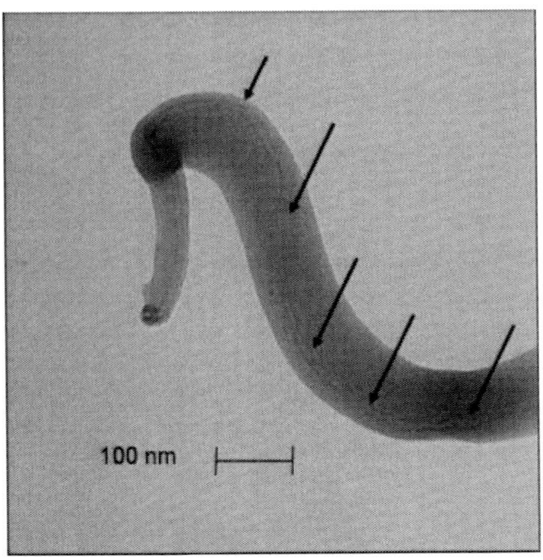

Figure 39. Electrospun polylactide nanofiber containing a multiwalled carbon nanotube. Note the alignment of the nanotube within the fiber and the lack of aggregating of the nanotube. (McCullen et al. Reprinted with permission from John Wiley and Sons).

3.4. Tendon and Ligament

Tendon (connecting muscle and bone) and ligament (connecting bone to bone) tissues are compositionally, structurally, and mechanically similar. Both tissues experience mechanical loads primarily in one direction and their extracellular matrix (mostly type I collagen) has a uniaxially aligned structure, leading to highly anisotropic mechanical properties (W. J. Li et al.). One clinical challenge is that tendons and ligaments have a low regenerative capacity due to their high extracellular matrix density, complex collagen organization, and low vascularity (Vunjak-Novakovic et al.). Thus, their scar-mediated healing response and inability to regenerate has led to the investigation of tissue engineering approaches to replace these damaged or diseased tendons and ligaments (Liu, Thomopoulos and Xia). Historically, braided and knitted fabrics have been applied as scaffolds for tendon/ligament repair. Due to their poor performance in cell adhesion, mass transport, and cell infiltration, it has been difficult to achieve functional tendon/ligament regeneration. Electrospun fibrous scaffolds have recently been utilized in an attempt to improve the biocompatibility and promote functional tendon/ligament regeneration.

3.4.2. Role of Electrospun Fiber Morphology in Tendon/Ligament Regeneration

With the importance of fiber morphology in tendon and ligament organization in mind, James et al. utilized electrospinning to create scaffolds composed of fiber diameters matching the structure of the native extracellular matrix. In this study, they were able to electrospin poly(lactic-co-glycolic acid) fibers in the biomimicking nanometer range and showed that these fibrous structures supported an increased human tendon fibroblast proliferation and extracellular matrix deposition *in vitro* (James et al.). Similarly, Cardwell et al. conducted a study in which electrospun scaffolds were fabricated from a variety of polymers and at three different fiber diameters (random and aligned fiber orientations): small (< 1 μm), medium (1–2 μm) and large (> 2 μm). The overall results demonstrated that an electrospun fiber diameter of less than 1 μm supported improved mesenchymal stem cell proliferation using C3H10T1/2 model stem cells and that fiber diameter was more of a factor in cell behaviors than alignment. After two weeks culture, the gene expression of collagen 1α1 and decorin was increased on all scaffolds compared to film controls. More importantly, the expression of the tendon and ligament transcription factor scleraxis was suppressed on all electrospun scaffolds relative to spin-coated control films, but the expression on the large-diameter fiber scaffolds was consistently greater than on the medium-diameter fiber scaffolds (Cardwell et al.).

Yin et al. studied the regulation of tendon stem/progenitor cell differentiation directed by the alignment of electrospun fibers. In this study, they used random and aligned poly(L-lactic acid) fibers. The human tendon stem/progenitor cells were cultured on the electrospun scaffolds to investigate the effect of electrospun matrix orientation. The expression of tendon-specific genes was significantly higher in human tendon stem/progenitor cells growing on aligned fibers as compared to those on randomly oriented fibers in both normal and osteogenic media. In addition, alkaline phosphatase activity and alizarin red staining illustrated that the randomly oriented scaffolds induced osteogenesis, while the aligned scaffold hindered the process (Yin et al. 2010). The results of their *in vivo* experiment showed that the aligned fibers induced the formation of spindle-shaped cells and tendon-like tissue. Similarly, C. H. Lee et al. demonstrated that scaffolds consisting of aligned electrospun polyurethane fibers guided human ligament fibroblasts to elongate and align in the longitudinal direction, which also provided the mechanical environment those fibroblasts encounter *in vivo*. Surrao et al. fabricated wavy, crimp-like poly(L-lactide-co-D,L-lactide) fibrous structures via electrospinning and subsequent heating treatment. Upon the removal of electrospun fiber scaffolds from the mandrel and placement into a heated aqueous environment overnight, they were able to produce crimp-like fiber scaffolds. The crimp-like pattern had an amplitude and wavelength that was similar to that observed in the collagen fibers of the anterior cruciate ligament. The crimp-like pattern appeared to aid the deposition of extracellular matrix at the surface of the fibrous scaffold by the seeded fibroblasts, facilitating fascicle formation. Additionally, cells and the presence of collagen accumulation were observed at the surface and interior regions of the constructs.

Xu et al. fabricated poly(L-lactide-co-ε-caprolactone)/collagen nanoyarn as a novel, three-dimensional, macro-porous, aligned scaffold for tendon tissue engineering using a water bath collector during electrospinning. They then cultured tendon cells on the yarn to determine their capacity for tendon tissue engineering. The tendon cells on the nanoyarn exhibited native morphology and improved cellular proliferation on the nanoyarn scaffold compared with cells cultured on the random and aligned electrospun fibers scaffolds. Moreover, the spatial structure (macro-pores) of the nanoyarn scaffold was beneficial for cell infiltration. In addition, this scaffold exhibited desirable mechanical properties for tendon tissue regeneration and replacement.

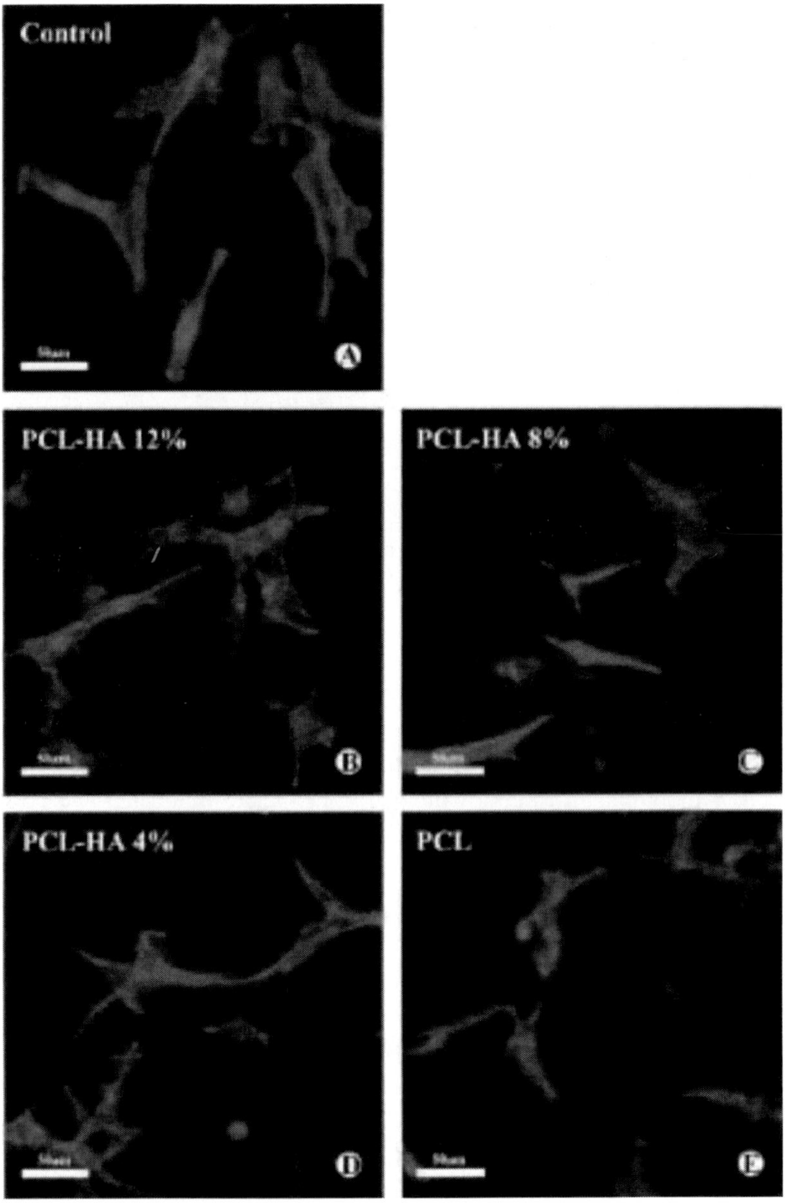

Figure 40. Images of double fluorescence stained (F-actin in red, nucleus in blue) cells cultured on the surface of (A) tissue culture plastic control, (B) polycaprolactone with 12% HA membrane, (C) polycaprolactone with 8% HA membrane, (D) polycaprolactone with 12% HA membrane, (E) polycaprolactone membrane with no HA incorporation. Scale bar is 50 μm. (S. Liu et al. Reprinted with permission from the American Chemical Society).

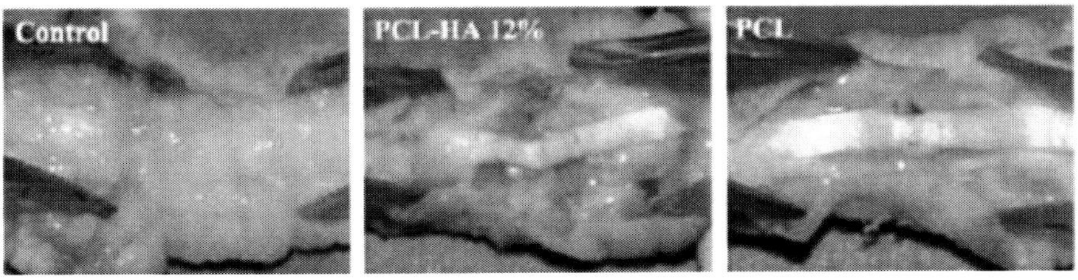

Figure 41. Images of chicken flexor digitorum profundus tendon model after 21 days post-implantation. (A) Non-treatment control, (B) bilayer HA-polycaprolactone/ polycaprolactone membrane treatment, (C) plain polycaprolactone scaffold. Adhesion was significantly lower in the HA-polycaprolactone/polycaprolactone bilayer treatment as compared to the polycaprolactone and non-treatment control. (S. Liu et al. Reprinted with permission from the American Chemical Society).

S. Liu et al. developed a multilayer biomimetic tendon sheath replacement for preventing adhesions during tendon replacement. A bilayer membrane was fabricated by electrospinning hyaluronic acid (HA)-embedded polycaprolactone fibers onto a membrane of non-HA-laden polycaprolactone fibers. These bilayer membranes were found to release HA over a 10 day period, with lower HA loading studies showed better adherence and distribution on the inner HA membrane side than the outer polycaprolactone side (Figure 40). Upon implantation in chicken flexor digitorum profundus tendon repair model, the bilayer membrane-repaired models exhibited significantly less peritendinous fibrous adhesion than non-treated and polycaprolactone controls (Figure 41). Biomechanical analysis of these models indicated that the work of flexion was significantly lower in the bilayer sheath than the polycaprolactone control and non-treatment control. These results indicated improved healing with the HA/polycaprolactone bilayer treatment, demonstrating its efficacy as biomimetic tendon sheath (S. Liu et al.).

3.4.3. Role of Mechanical Stimulation

Mechanical stimulation and tissue preconditioning has been identified as a critical component in engineering load-bearing tissues such as muscle, bone, cartilage, and ligament/tendon. It is clear that without the appropriate biomechanical cues, the new tissue formed will lack the necessary extracellular matrix quantity and organization to be sufficient for functional, load-bearing capacity (Murray, Vavken and Fleming). Thus, a great deal of current research focuses on the use of cyclic mechanical loading to direct fibroblast alignment and guide cell morphology to an elongated, spindle-like shape during tendon and ligament tissue engineering. In addition, the use of mechanical stimulation also increases cellular proliferation, collagen synthesis, and growth factor expression to benefit the ability to regenerate functional tissues (Benhardt and Cosgriff-Hernandez).

In one example of using mechanical stimulation in tendon regeneration, Barber et al. fabricated braided electrospun poly(L-lactic acid) scaffolds with highly oriented fibers into tendon scaffolds. Results from the use of these scaffolds along with applied mechanical stimulation showed that they were able to promote the differentiation of human mesenchymal stem cells. Moreover, their results indicated that the braided scaffolds enhanced the mechanobiological properties of electrospun fibers, along with tenogenic growth factors supplementing the media, allowed for the expression of key tenocyte markers. Thus, it was proposed that the braided nanofibrous scaffolds could act as scaffolding capable of supporting cell expansion and differentiation through sequential culture in growth medium first, followed by differentiation medium and cyclic tensile stimulation.

The effect of mechanical stimulation on the nanoyarn tendon scaffolds was also studied by Xu et al. in 2014. The results of that study showed that mechanical stimulation can enhance cell proliferation and subsequent tenogeneic differentiation of tendon-derived stem cells *in vitro*. Additionally, dynamically cultured tendon-derived stem cells on poly(L-lactic acid-co-polycaprolactone)/collagen constructs significantly promoted the repair of injured rabbit patellar tendons with excellent resulting mechanical properties, which resulted from enhanced collagen production and tendon-related protein expressions. Surrao et al. investigated the influence of mechanical stimuli in the tissue production of fibroblasts cultured on the crimp-like scaffold. The results of that study indicated that the long-term, dynamic mechanical stimulation of fibroblasts seeded on crimp-like fiber scaffolds at 10% strain amplitude resulted in significantly up-regulated collagen accumulation and down-regulated sulfated proteoglycan accumulation. An image of a typical mechanical stimulation bioreactor is shown in Figure 42. In addition, the fibroblasts seeded on the dynamically stimulated crimp-like fiber scaffolds formed bundles that resembled fascicles. Multi-type dynamic culture methods were applied to the highly aligned fiber scaffolds for ligament regeneration (Mizutani et al.). More specifically, this study used a highly oriented fiber scaffold composed of electrospun elastin and collagen. In the study, these scaffolds were seeded with human periodontal ligament fibroblasts. They were then cultured using a combination of pressure and twist stimulation in order to examine the influence of mechanical force on the functional maintenance of ligament cells and their differentiation to osteoblast-like cells. The results indicated that a pressure stimulation and the presence of elastin A led to an up regulation in the expression of alkaline phosphatase and promoted the osteogenic differentiation of ligament cells. The twist stimulation upregulated the expression of type III collagen. Furthermore, the combination of pressure and twist stimulation promoted the expression of type III collagen and alkaline phosphatase depending on the portion of scaffold.

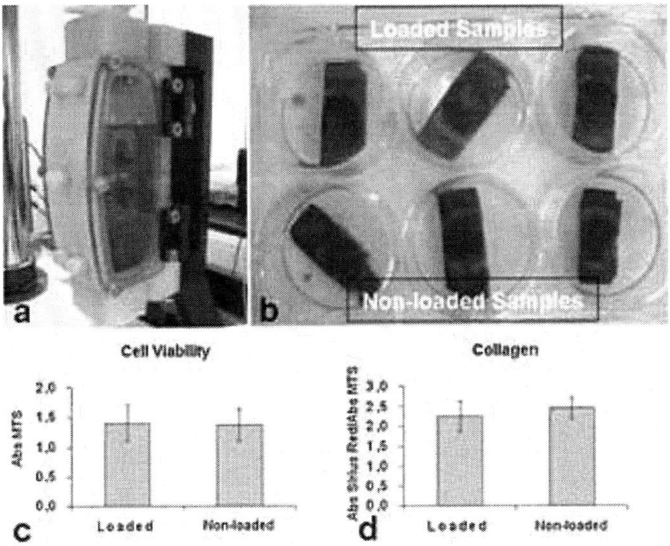

Figure 42. Image of a typical mechanical stimulation bioreactor. (Puppi et al. Reprinted with permission from John Wiley and Sons).

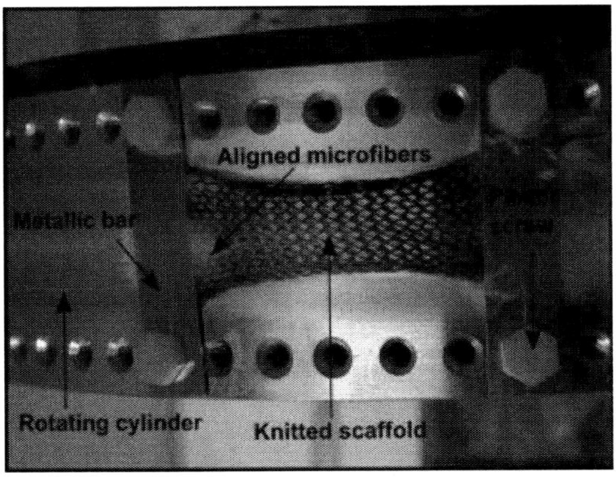

Figure 43. Setup for the creation of hybrid electrospun/knit scaffolds. The knitted scaffold was placed on a rotating drum onto which fibers were electrospun to create a porous layer. (C. Vaquette et al. 2010. Reprinted with permission of John Wiley and Sons).

3.4.4. Electrospun Hybrid Fiber for Scaffolds

Electrospun fibrous scaffolds always possess poor mechanical properties relative to the native tissue, even if fabricated into parallel structures. However, the tendon and ligament tissue substitutes have the demand for high strength. Thus, hybrid scaffolds of fabrics such as braided, knitted fabrics have been incorporated with electrospun fibers. It should be noted that the braided (H. H. Liu et al. 2014, Hu and Huang) and knitted (X. Chen et al., X. Wang et al., J. L. Chen et al.) fabrics have been applied in tendon regeneration for many years. However, the limitations of textile techniques and the structure of fabrics, nutrient transmission, poor cell seeding, infiltration, and matrix production have been major downsides of their potential use (H. Liu et al.).

In addition, a controlled and homogeneous cell seeding is very difficult to achieve. To overcome these limitations, different techniques have been developed, such as sponge formation (X. Chen et al., Shen et al., J.L Chen et al.) and electrospun fibers (Sahoo, Cho-Hong and Siew-Lok; Sahoo, Toh and Goh; Funakoshi et al., Vaquette et al., Sahoo et al.).

In a study by Sahoo et al., they electrospun poly(D, L-lactide-co-glycolide) fibers onto a knitted poly (D, L-lactide-co-glycolide) scaffold to provide a large surface for cell attachment. They also prepared knitted poly(D, L-lactide-co-glycolide) scaffolds coated with a type 1 collagen and a polycaprolactone film (Sahoo, Cho-Hong and Siew-Lok). These hybrid scaffolds possessed good mechanical properties and were easily seeded with cells. The modification techniques of coating with nanofibers or collagen were particularly effective in aiding cell attachment, growth and proliferation, thus creating the potential for use in tendon and ligament tissue engineering.

In another study, aligned poly(L-lactic-co-e-caprolactone) microfibers were electrospun onto knitted poly(L-lactic-co-ε-caprolactone) scaffolds (Figure 43) (Vaquette et al.). Upon seeding of these structures with rat bone marrow-derived mesenchymal stem cells, the microfiber architecture induced a high level of extracellular matrix secretion (Figure 44). It was found that cells primarily produced collagen type I and type III, two main components found in ligaments. After fourteen days of culture, collagen type III started to form a fibrous network. A study by Teh et al. utilized a composite scaffold composed of knitted silk fibroin and aligned silk fibroin electrospun fibers to fabricate a three-dimensional hybrid scaffold system. In this system, the knitted silk fibroin contributed to the mechanical robustness while the aligned silk fibers acted as the initial extracellular matrix to provide environmental cues for positive cellular response. The mesenchymal stem cells seeded on the aligned fibrous hybrid scaffolds were shown to be proliferative and aligned along the electrospun silk fibers, forming oriented spindle-shaped morphology and produced an aligned extracellular matrix network. The expression and production of ligament-related proteins was also increased as compared to hybrid silk scaffolds with randomly arranged fibers, indicating differentiation cues for ligament fibroblasts present in the aligned hybrid silk scaffolds.

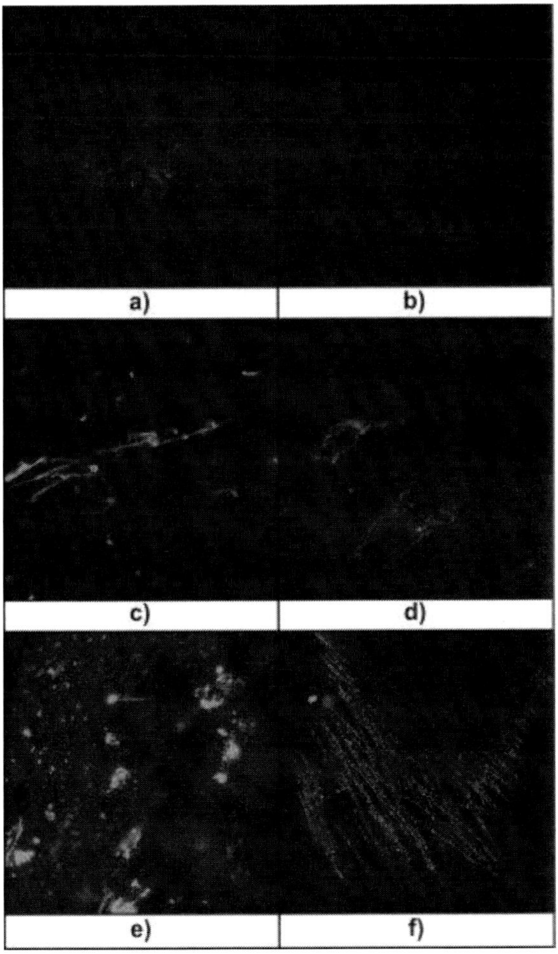

Figure 44. Rat mesenchymal stem cells seeded on electrospun/knit hybrid scaffolds of poly(L-lactic-co-e-caprolactone). (A and B) Cells seeded in tissue culture plastic, (C and D) cells seeded on hybrid scaffold after 7 days culture, (E and F) Cells seeded on hybrid scaffold after 14 days culture. (A, C, and E) are stained for collagen type I, and (B,D, and F) are stained for collagen type III. Note the increase in collagen production over time. (C. Vaquette et al. 2010. Reprinted with permission from John Wiley and Sons).

As an example of a drug delivery scaffolding, Sahoo et al. encapsulated basic fibroblast growth factor in electrospun poly(L-lactic-co-glycolic acid) fibers which were deposited onto the surface of a knitted silk scaffold. The released basic fibroblast growth factor stimulated initial proliferation and subsequent tenogeneic differentiation of the seeded bone marrow-derived mesenchymal stem cells, which enhanced their collagen production and resulted in stronger constructs. The bio-hybrid scaffold system not only facilitated mesenchymal progenitor cell attachment and promoted cell proliferation, with cells growing both on ultrafine poly(L-lactic-co-glycolic acid) fibers and silk micro-fibres, but also stimulated tenogeneic differentiation of seeded mesenchymal progenitor cells. It should be noted that the upregulated gene expression of ligament/tendon-specific extracellular matrix proteins and increased collagen production likely contributed to enhancing mechanical properties of these constructs.

3.5. Electrospun Skin Substitutes

Skin, the largest organ of the body, consists of two tissue layers: a keratinized stratified epidermis and an underlying layer of collagen-rich connective tissue known as the dermis. Injured or damaged skin has the innate ability to repair itself. However, if the degree of injury is large enough, this capacity to regenerate is severely compromised. The use of xenografts, allografts, and autografts can aid in the healing process, however, due to antigenicity or limited donor sites these skin substitutes cannot always achieve the task of providing a complete barrier coverage. This creates a need for an alternative, which is where skin tissue engineering comes in. Throughout the past several decades, a variety of materials have been used in the manufacture of skin tissue engineering scaffolds, including cellulose derivatives, poly(vinyl alcohol), silicone, collagen, and composite materials such as collagen/chitosan, polycaprolactone/collagen, and poly(L-lactic acid)/collagen hybrid scaffolds.

Electrospun collagen fibers have been shown to promote cell growth and penetration of cells into the engineered matrix. Jin et al. have electrospun collagen scaffolds for epidermal differentiation of mesenchymal stem cells. In that study, early and intermediate epidermal differentiation markers were expressed and the regional differentiation was greatly reduced using a substrate of electrospun nanofibers as opposed to tissue culture plastic. In another study, electrospun gelatin scaffolds were used to study the effect of fiber density on dermal-epidermal skin regeneration (Powell and Boyce). Electrospun scaffolds with varying fiber diameter, inter-fiber distance, and porosity were fabricated by altering the concentration of the electrospinning solution. Inter-fiber distances greater than 5.5 μm allowed deeper penetration of human dermal fibroblasts into the scaffold, whereas cells in scaffolds with more densely packed fibers were able to infiltrate only the upper regions. Inter-fiber distances between 5 and 10 μm appear to yield the most favorable skin substitute *in vitro*, demonstrating high cell viability, optimal cell organization, and excellent barrier formation.

Zhu et al. developed an electrospun dermal substitute with a porosity as high as 92.4% through which fibroblasts could migrate and populate. Higher human dermal fibroblast viability, collagen deposition, cell migration, and infiltration inside the scaffold were achieved by the fibrous scaffold with the higher porosity. Kumbar et al. conducted a study that examined the effect of fiber diameter of electrospun poly(lactic-co-glycolic acid) on their potential application for treating severe skin loss. In their study, poly(lactic-co-glycolic acid) fiber diameters of 150–225, 200– 300, 250–467, 500–900, 600–1200, 2500–3000 and 3250–6000 nm were obtained via electrospinning. Seeded human dermal fibroblasts showed a well spread morphology and significant growth on fiber matrices in the 350–1100 nm diameter range. Collagen type III gene expression was also significantly up-regulated in fibroblasts seeded on scaffolds with fiber diameters in the range of 350–1100 nm.

Cui et al. fabricated composite electrospun fibrous scaffolds composed of poly(D,L-lactic acid) and poly(ethylene glycol) for skin tissue engineering. Seeded human dermal fibroblasts interacted and integrated well with the surrounding fibers containing 20 and 30% poly(ethylene glycol), which provided a significantly improved environment for biological activities of the dermal fibroblast than simply electrospun poly(D,L-lactic acid) scaffolds. It was also demonstrated that electrospun scaffolds containing 30% poly(ethylene glycol) exhibited the most balanced properties of moderately hydrophilic surface, minimal dimensional shrinkage, adaptable bulk biodegradation profile, and the enhancement of cell penetration within the fibrous scaffolds. Collagen/chitosan (Ma, 2003), chitin (Noh et al.), polycaprolactone/collagen composite, gelatin/polycaprolactone composite (Chong et al.) fibers have also been electrospun and applied in the regeneration of skin with varying results.

REFERENCES

Badami, A.S., et al. "Effect of Fiber Diameter on Spreading, Proliferation, and Differentiation of Osteoblastic Cells on Electrospun Poly(Lactic Acid) Substrates." *Biomaterials* 27 (2006): 596-606.

Barber, J.G., et al. "Braided Nanofibrous Scaffold for Tendon and Ligament Tissue Engineering." *Tissue Engineering Part A* 19 (2013): 1265-74.

Benhardt, H.A., and E.M. Cosgriff-Hernandez. "The Role of Mechanical Loading in Ligament Tissue Engineering." *Tissue Engineering Part B: Reviews* 15 (2009): 467-75.

Bhang, S.H., et al. "Enhanced Nerve Growth Factor Efficiency in Neural Cell Culture by Immobilization on the Culture Substrate." *Biochemical and Biophysical Research Communications* 382 (2009): 315-20.

Boland, E.D., et al. "Electrospinning Collagen and Elastin: Preliminary Vascular Tissue Engineering." *Frontiers in Bioscience* 9 (2004): 1422-32.

Boland, E.D., et al. "Electrospinning Polydioxanone for Biomedical Applications." *Acta Biomaterialia* 1 (2005): 115-123.

Cao, H., T. Liu, and S.Y. Chew. "The Application of Nanofibrous Scaffolds in Neural Tissue Engineering." *Advanced Drug Delivery Reviews* 61 (2009): 1055-64.

Cardwell, R.D., L.A. Dahlgren, and A.S. Goldstein. "Electrospun Fibre Diameter, Not Alignment, Affects Mesenchymal Stem Cell Differentiation into the Tendon/Ligament Lineage." *Journal of Tissue Engineering and Regenerative Medicine* 8 (2012): 937-45.

Chan, C. K., et al. "Early Adhesive Behavior of Bone-Marrow-Derived Mesenchymal Stem Cells on Collagen Electrospun Fibers." *Biomedical Materials* 4 (2009): 035006.

Chen, F., Huang, P., and X. Mo. "Electrospinning of Heparin Encapsulated P (LLA-Cl) Core/Shell Nanofibers." *Nano Biomedicine and Engineering* 2 (2010): 84-90.

Chen, J.L., et al. "Efficacy of hESC-MSCs in Knitted Silk-Collagen Scaffold for Tendon Tissue Engineering and Their Roles." *Biomaterials* 31 (2010): 9438-51.

Chen, X., et al. "Ligament Regeneration Using a Knitted Silk Scaffold Combined with Collagen Matrix." *Biomaterials* 29 (2008): 3683-92.

Chiu, J.B., et al. " Functionalization of Poly(L-lactide) Nanofibrous Scaffolds with Bioactive Collagen Molecules." *Journal of Biomedical Materials Research Part A* 83 (2007): 1117-27.

Cho, Y.I., et al. "Nerve Growth Factor (NGF)-Conjugated Electrospun Nanostructures with Topographical Cues for Neuronal Differentiation of Mesenchymal Stem Cells." *Acta Biomaterilaia* 6 (2010): 4725-33.

Chong, E.J., et al. "Evaluation of Electrospun PCL/Gelatin Nanofibrous Scaffold for Wound Healing and Layered Dermal Reconstitution." *Acta Biomaterilia* 3 (2007): 321-30.

Chronakis, I.S., Grapenson, S., and Jakob, A. "Conductive Polypyrrole Nanofibers Via Electrospinning: Electrical and Morphological Properties." *Polymer* 47 (2006): 1597-603.

Cleary, M.A., et al. "Vascular Tissue Engineering: The Next Generation." *Trends in Molecular Medicine* 18 (2012): 394-404.

Cui, W., et al. "Evaluation of Electrospun Fibrous Scaffolds of Poly(DL-Lactide) and Poly(Ethylene Glycol) for Skin Tissue Engineering." *Materials Science and Engineering: C* 29 (2009): 1869-76.

de Valence, S., et al. "Long Term Performance of Polycaprolactone Vascular Grafts in a Rat Abdominal Aorta Replacement Model." *Biomaterials* 33 (2012): 38-47.

Detta, N., et al. "Novel Electrospun Polyurethane/Gelatin Composite Meshes for Vascular Grafts." *Journal of Materila Science: Materilas in Medicine* 21 (2010): 1761-9.

Du, F., et al. "Gradient Nanofibrous Chitosan/Poly Varepsilon-Caprolactone Scaffolds as Extracellular Microenvironments for Vascular Tissue Engineering." *Biomaterials* 33 (2012): 762-70.

Funakoshi, T., et al. "Novel Chitosan-Based Hyaluronan Hybrid Polymer Fibers as a Scaffold in Ligament Tissue Engineering." *Journal of Biomedical Materials Research Part A* 74 (2005): 338-46.

Ghasemi-Mobarakeh, L., et al. "Electrospun Poly(Epsilon-Caprolactone)/Gelatin Nanofibrous Scaffolds for Nerve Tissue Engineering." *Biomaterials* 29 (2008): 4532-9.

Ghasemi-Mobarakeh, L., et al. "Electrical Stimulation of Nerve Cells Using Conductive Nanofibrous Scaffolds for Nerve Tissue Engineering." *Tissue Engineering Part A* 15 (2009): 3605-19.

Hashi, C.K., et al. "Antithrombogenic Modification of Small-Diameter Microfibrous Vascular Grafts." *Arteriosclerosis, Thrombosis, and Vascular Biology* 30 (2010): 1621-7.

He, C., et al. "Physical-Chemical Properties and *in vitro* Biocompatibility Assessment of Spider Silk, Collagen and Polyurethane Nanofiber Scaffolds for Vascular Tissue Engineering." *Nano Biomedicine and Engineering* 1 (2009): 80-88.

He, W., et al. "Tubular Nanofiber Scaffolds for Tissue Engineered Small-Diameter Vascular Grafts," *Journal of Biomedical Materials Research Part A* 90 (2009): 205-16.

He, W., et al. "Fabrication and Endothelialization of Collagen-Blended Biodegradable Polymer Nanofibers: Potential Vascular Graft for Blood Vessel Tissue Engineering." *Tissue Engineering* 11 (2005): 1574-88.

Hong, Y., et al. "A Small Diameter, Fibrous Vascular Conduit Generated from a Poly(Ester Urethane)Urea and Phospholipid Polymer Blend." *Biomaterials* 30 (2009): 2457-67.

Hu, W. and Z.M. Huang. "Biocompatibility of Braided Poly(L-Lactic Acid) Nanofiber Wires Applied as Tissue Sutures." *Polymer International* 59 (2010): 92-99.

James, R., et al. "Electrospun Nanofibrous Scaffolds for Engineering Soft Connective Tissues." *Methods in Molecular Biology* 726 (2011): 243-58.

Jang, J.H., Castano, O., and Kim, H.W. "Electrospun Materials as Potential Platforms for Bone Tissue Engineering." *Advanced Drug Delivery Reviews* 61 (2009): 1065-83.

Jin, G., Prabhakaran, M.P., and Ramakrishna, S. "Stem Cell Differentiation to Epidermal Lineages on Electrospun Nanofibrous Substrates for Skin Tissue Engineering." *Acta Biomaterialia* 7 (2011): 3113-22.

Karageorgiou, V., and Kaplan, D. "Porosity of 3D Biomaterial Scaffolds and Osteogenesis." *Biomaterials* 26 (2005): 5474-91.

Kim, H.W., Lee, H.H., and Knowles, J.C. "Electrospinning biomedical nanocomposite fibers of hydroxyapatite/poly(lactic acid) for bone regeneration." *Journal of Biomedical Materials Research Part A* 79A (2006): 643-649.

Kim, Y.T., et al. "The Role of Aligned Polymer Fiber-Based Constructs in the Bridging of Long Peripheral Nerve Gaps." *Biomaterials* 29 (2008): 3117-27.

Kumbar, S.G., et al. "Electrospun Poly(Lactic Acid-Co-Glycolic Acid) Scaffolds for Skin Tissue Engineering." *Biomaterials* 29 (2008): 4100-7.

Lee, A.C., et al. "Controlled Release of Nerve Growth Factor Enhances Sciatic Nerve Regeneration." *Experimental Neurology* 184 (2003): 295-303.

Lee, C.H., et al. "Nanofiber Alignment and Direction of Mechanical Strain Affect the ECM Production of Human ACL Fibroblast." *Biomaterials* 26 (2005): 1261-70.

Lee, S.J., et al. "Development of a Composite Vascular Scaffolding System That Withstands Physiological Vascular Conditions." *Biomaterials* 29 (2008): 2891-8.

Li, C., et al. "Electrospun Silk-BMP-2 Scaffolds for Bone Tissue Engineering." *Biomaterials* 27 (2006): 3115-24.

Li, M., et al. "Electrospun protein fibers as matrices for tissue engineering." *Biomaterials* 26 (2005): 5999-6008.

Li, W.J., et al. "Engineering Controllable Anisotropy in Electrospun Biodegradable Nanofibrous Scaffolds for Musculoskeletal Tissue Engineering." *Journal of Biomechics* 40 (2007): 1686-93.

Li, X., et al. "Coating Electrospun Poly (E-Caprolactone) Fibers with Gelatin and Calcium Phosphate and Their Use as Biomimetic Scaffolds for Bone Tissue Engineering." *Langmuir* 24 (2008): 14145-50.

Liu, H., et al. "The Interaction between a Combined Knitted Silk Scaffold and Microporous Silk Sponge with Human Mesenchymal Stem Cells for Ligament Tissue Engineering." *Biomaterials* 29 (2008): 662-74.

Liu, J.J., et al. "Peripheral Nerve Regeneration Using Composite Poly(Lactic Acid-Caprolactone)/Nerve Growth Factor Conduits Prepared by Coaxial Electrospinning." *Journal of Biomedical Materials Research Part A* 96 (2011): 13-20.

Liu, S. et al. "Biomimetic Sheath Membrane via Electrospinning for Antiadhesion of Repaired Tendon." *Biomacromolecules* 13 (2012): 3611-3619.

Liu, T., et al. "Photochemical Crosslinked Electrospun Collagen Nanofibers: Synthesis, Characterization and Neural Stem Cell Interactions." *Journal of Biomedical Materials Research Part A* 95 (2010): 276-82.

Liu, W., Thomopoulos, S. and Xia, Y. "Electrospun Nanofibers for Regenerative Medicine." *Advanced Healthcare Materials* 1 (2012): 10-25.

Ma, L. "Collagen/Chitosan Porous Scaffolds with Improved Biostability for Skin Tissue Engineering." *Biomaterials* 24 (2003): 4833-41.

Ma, L. et al. "Trap Effect of Three-Dimensional Fibers Network for High Efficient Cancer-Cell Capture." *Advanced Healthcare Materials* 4 (2015): 838-843

Martins, A., et al. "Surface Modification of Electrospun Polycaprolactone Nanofiber meshes by Plasma Treatment to Enhance Biological Performance." *Small* 5 (2009) 1195-1206.

Madurantakam, P.A. et al. "Evaluation of Biological Activity of Bone Morphogenetic Proteins on Exposure to Commonly Used Electrospinning Solvents." *Journal of Bioactive and Compatible Polymers: Biomedical Applications* 28 (2013): 154-166.

Matthews, J.A. et al. "Electrospinning of Collagen Type II: A Feasibility Study." *Journal of Bioactive and Compatible Polymers* 18 (2003): 125-34.

McClure, M.J., et al. "The Use of Air-Flow Impedance to Control Fiber Deposition Patterns During Electrospinning." *Biomaterials* 33 (2012): 771-9.

McClure, M.J., et al. "Electrospinning Aligned and Random Polydioxanone- and Polycaprolactone- Silk Fibroin Blended Scaffolds: Geometry for a Vascular Matrix." *Biomedical Materials* 4 (2009): 055010.

McCullen et al. "Development, Optimization, and Characterization of Electrospun Poly(lactic acid) Nanofibers Containing Multi-Walled Carbon Nanotubes." *Journal of Applied Polymer Science* 105 (2007): 1668-1678.

Min, B.M., et al. "Chitin and Chitosan Nanofibers: Electrospinning of Chitin and Deacetylation of Chitin Nanofibers." *Polymer* 45 (2004): 7137-42.

Mizutani, N., et al. "Multiple-Type Dynamic Culture of Highly Oriented Fiber Scaffold for Ligament Regeneration." *Journal of Artificial Organs* 16 (2013): 49-58.

Murray, M.M., Vavken, P. and Fleming, B.C. "The ACL Handbook." (2013).

Nandakumar, A., et al. "Surface Modification of Electrospun Fibre Meshes by Oxygen Plasma for Bone Regeneration." *Biofabrication* 5 (2013): 015006.

Nie, H., and Wang, C.H.. "Fabrication and Characterization of PLGA/HAP Composite Scaffolds for Delivery of BMP-2 Plasmid DNA." *Journal of Controlled Release* 120 (2007): 111-21.

Nisbet, D.R., et al. "Interaction of Embryonic Cortical Neurons on Nanofibrous Scaffolds for Neural Tissue Engineering." *Journal of Neural Engineering* 4 (2007): 35-41.

Noh, H.K., et al. "Electrospinning of Chitin Nanofibers: Degradation Behavior and Cellular Response to Normal Human Keratinocytes and Fibroblasts." *Biomaterials* 27 (2006): 3934-44.

Nottelet, B., et al. "Factorial Design Optimization and *in vivo* Feasibility of Poly(Epsilon-Caprolactone) Micro- and Nanofiber-Based Small Diameter Vascular Grafts." *Journal of Biomedical Materials Research Part A* 89 (2009): 865-75.

Park, S.Y., et al. "Electrospun Silk Fibroin Scaffolds with Macropores for Bone Regeneration: An *in vitro* and *In vivo* Study." *Tissue Engineering Part A* 16 (2010): 1271-79.

Phipps, M.C., et al. "Increasing the Pore Sizes of Bone-Mimetic Electrospun Scaffolds Comprised of Polycaprolactone, Collagen I and Hydroxyapatite to Enhance Cell Infiltration." *Biomaterials* 33 (2012): 524-34.

Powell, H.M., and Boyce, S.T. "Fiber Density of Electrospun Gelatin Scaffolds Regulates Morphogenesis of Dermal-Epidermal Skin Substitutes." *Journal of Biomedical Materials Research Part A* 84 (2008): 1078-86.

Prabhakaran, M.P., et al. "Electrospun Conducting Polymer Nanofibers and Electrical Stimulation of Nerve Stem Cells." *Journal of Bioscience and Bioengineering* 112 (2011): 501-7.

Prabhakaran, M.P., et al. "Surface Modified Electrospun Nanofibrous Scaffolds for Nerve Tissue Engineering." *Nanotechnology* 19 (2008): 455102.

Prabhakaran, M.P., et al. "Electrospun Biocomposite Nanofibrous Scaffolds for Neural Tissue Engineering." *Tissue Engineering Part A* 14 (2008): 1787-97.

Puppi, D. et al. "Development of Electrospun Three-Arm Star Poly(ε-caprolactone) for Tissue Engineering Applications." *Macromolecular Bioscience* 10 (2010): 887-897.

Rivers, T.J., Hudson, T.W., and Schmidt, C.E.. "Synthesis of a Novel, Biodegradable Electrically Conducting Polymer for Biomedical Applications." *Advanced Functional Materials* 12 (2002): 33-37.

Roh, J.D., et al. "Small-Diameter Biodegradable Scaffolds for Functional Vascular Tissue Engineering in the Mouse Model." *Biomaterials* 29 (2008): 1454-63.

Sahoo, S., et al. "Characterization of a Novel Polymeric Scaffold for Potential Application in Tendon/Ligament Tissue Engineering." *Tissue Engineering* 12 (2006): 91-99.

Sahoo, S., Cho-Hong, J.G., and Siew-Lok, T. "Development of Hybrid Polymer Scaffolds for Potential Applications in Ligament and Tendon Tissue Engineering." *Biomedical Materials* 2 (2007): 169-73.

Sahoo, S., Toh, S.L., and Goh, J.C. "A bFGF-Releasing Silk/PLGA-based Biohybrid Scaffold for Ligament/Tendon Tissue Engineering Using Mesenchymal Progenitor Cells." *Biomaterials* 31 (2010): 2990-8.

Schnell, E., et al. "Guidance of Glial Cell Migration and Axonal Growth on Electrospun Nanofibers of Poly-Epsilon-Caprolactone and a Collagen/Poly-Epsilon-Caprolactone Blend." *Biomaterials* 28 (2007): 3012-25.

Shao, S., et al. "Osteoblast Function on Electrically Conductive Electrospun PLA/Mwcnts Nanofibers." *Biomaterials* 32 (2011): 2821-33.

Shen, W., et al. "The Effect of Incorporation of Exogenous Stromal Cell-Derived Factor-1 Alpha within a Knitted Silk-Collagen Sponge Scaffold on Tendon Regeneration." *Biomaterials* 31 (2010): 7239-49.

Shin, M., Yoshimoto, H. and Vacanti, J.P.. "In vivo Bone Tissue Engineering Using Mesenchymal Stem Cells on a Novel Electrospun Nanofibrous Scaffold" Tissue Engineering 10 (2004): 33-41.

Soletti, L., et al. "*In vivo* Performance of a Phospholipid-Coated Bioerodable Elastomeric Graft for Small-Diameter Vascular Applications." *Journal of Biomedical Materials Research Part A* 96 (2011): 436-48.

Song, W., et al. "Electrospun Polyvinyl Alcohol-Collagen-Hydroxyapatite Nanofibers: A Biomimetic Extracellular Matrix for Osteoblastic Cells." *Nanotechnology* 23 (2012): 115101.

Stitzel, J.D., et al. "Arterial Smooth Muscle Cell Proliferation on a Novel Biomimicking, Biodegradable Vascular Graft Scaffold." *Journal of Biomaterials Applications* 16 (2001): 22-33.

Su, Y., et al. "Encapsulation and Controlled Release of Heparin from Electrospun Poly (L-Lactide-Co-E-Caprolactone) Nanofibers." *Journal of Biomaterials Science, Polymer Edition* 22 (2011): 165-77.

Surrao, D.C., et al. "A Crimp-Like Microarchitecture Improves Tissue Production in Fibrous Ligament Scaffolds in Response to Mechanical Stimuli." *Acta Biomaterialia* 8 (2012): 3704-13.

Surrao, D.C., Waldman, S.D., and Amsden, B.G. "Biomimetic Poly(Lactide) Based Fibrous Scaffolds for Ligament Tissue Engineering." *Acta Biomaterialia* 8 (2012): 3997-4006.

Teh, T.K., Toh, S.L., and Goh, J.C. "Aligned Hybrid Silk Scaffold for Enhanced Differentiation of Mesenchymal Stem Cells into Ligament Fibroblasts." *Tissue Engineering Part C Methods* 17 (2011): 687-703.

Tillman, B.W., et al. "The *In vivo* Stability of Electrospun Polycaprolactone-Collagen Scaffolds in Vascular Reconstruction." *Biomaterials* 30 (2009): 583-8.

Tillman, B., Kim, J., and Harbuzariu, A. "A Novel Depletion Approach Attenuates a Surge of CD34 Progenitors after Vascular Injury: Implications for Restenotic Vascular Disease." *Journal of Vascular Surgery* 56 (2012): 1824.

Valmikinathan, C.M., Defroda, S., and Yu, X. "Polycaprolactone and Bovine Serum Albumin Based Nanofibers for Controlled Release of Nerve Growth Factor." *Biomacromolecules* 10 (2009): 1084-89.

Vaquette, C., et al. "Aligned Poly(L-Lactic-Co-E-Caprolactone) Electrospun Microfibers and Knitted Structure: A Novel Composite Scaffold for Ligament Tissue Engineering." *Journal of Biomedical Materials Research Part A* 94 (2010): 1270-82.

Vaz, C.M., et al. "Design of Scaffolds for Blood Vessel Tissue Engineering Using a Multi-Layering Electrospinning Technique." *Acta Biomaterialia* 1 (2005): 575-82.

Vunjak-Novakovic, G., et al. "Tissue Engineering of Ligaments." *Annual Review of Biomedical Engineering* 6 (2004): 131-56.

Wang, C.Y., et al. "The Effect of Aligned Core–Shell Nanofibres Delivering NGF on the Promotion of Sciatic Nerve Regeneration." *Journal of Biomaterials Science, Polymer Edition* 23 (2012): 167-84.

Wang, G., et al. "Electrospun PLGA-Silk Fibroin-Collagen Nanofibrous Scaffolds for Nerve Tissue Engineering." *in vitro Cellular & Developmental Biology - Animal* 47 (2011): 234-40.

Wang, T., et al. "Fabrication and Characterization of Heparin-Grafted Poly-L-Lactic Acid-Chitosan Core-Shell Nanofibers Scaffold for Vascular Gasket." *ACS Appllied Material Interfaces* 5 (2013): 3757-63.

Wang, W., et al. "Enhanced Nerve Regeneration through a Bilayered Chitosan Tube: The Effect of Introduction of Glycine Spacer into the Cyigsr Sequence." *Journal of Biomedical Materials Research Part A* 85 (2008): 919-28.

Wang, W., et al. "Influences of Mechanical Properties and Permeability on Chitosan Nano/Microfiber Mesh Tubes as a Scaffold for Nerve Regeneration." *Journal of Biomedical Materials Research Part A* 84 (2008): 557-66.

Wang, W., et al. "Effects of Schwann Cell Alignment Along the Oriented Electrospun Chitosan Nanofibers on Nerve Regeneration." *Journal of Biomedical Materials Research Part A* 91 (2008): 994-1005.

Wang, X., et al. "The Roles of Knitted Mesh-Reinforced Collagen-Chitosan Hybrid Scaffold in the One-Step Repair of Full-Thickness Skin Defects in Rats." *Acta Biomaterialia* 9 (2013): 7822-32.

Willerth, S. M., and S. E. Sakiyama-Elbert. "Approaches to Neural Tissue Engineering Using Scaffolds for Drug Delivery." *Advanced Drug Delivery Reviews* 59 (2007): 325-38.

Xiang, P., et al. "Cytocompatibility of Electrospun Nanofiber Tubular Scaffolds for Small Diameter Tissue Engineering Blood Vessels." *International Journal of Biological Macromolecules* 49 (2011): 281-88.

Xie, J., et al. "Conductive Core-Sheath Nanofibers and Their Potential Application in Neural Tissue Engineering." *Advanced Functional Materials* 19 (2009): 2312-2318.

Xie, J., et al. "The Differentiation of Embryonic Stem Cells Seeded on Electrospun Nanofibers into Neural Lineages." *Biomaterials* 30 (2009): 354-62.

Xu, C.Y., et al. "Aligned Biodegradable Nanofibrous Structure: A Potential Scaffold for Blood Vessel Engineering." *Biomaterials* 25 (2004): 877-86.

Xu, Y., et al. “The Effect of Mechanical Stimulation on the Maturation of TDSCs-Poly(L-Lactide-Co-E-Caprolactone)/Collagen Scaffold Constructs for Tendon Tissue Engineering.” *Biomaterials* 35 (2014): 2760-72.

Xu, Y., et al. “Fabrication of Electrospun Poly(L-Lactide-Co-Varepsilon-Caprolactone)/Collagen Nanoyarn Network as a Novel, Three-Dimensional, Macroporous, Aligned Scaffold for Tendon Tissue Engineering.” *Tissue Enginerring Part C Methods* 19 (2013): 925-36.

Yang, F., et al. “Characterization of Neural Stem Cells on Electrospun Poly (L-Lactic Acid) Nanofibrous Scaffold.” *Journal of Biomaterials Science, Polymer Edition* 15 (2004): 1483-97.

Yang, F., et al. “Electrospinning of Nano/Micro Scale Poly(L-Lactic Acid) Aligned Fibers and Their Potential in Neural Tissue Engineering.” *Biomaterials* 26 (2005): 2603-10.

Yang, F., Wolke, J.G.C., and Jansen, J.A. “Biomimetic Calcium Phosphate Coating on Electrospun Poly(ε-Caprolactone) Scaffolds for Bone Tissue Engineering.” *Chemical Engineering Journal* 137 (2008): 154-61.

Yin, A., et al. “Electrospinning Collagen/Chitosan/Poly(L-Lactic Acid-Co-Epsilon-Caprolactone) to Form a Vascular Graft: Mechanical and Biological Characterization.” *Journal of Biomedical Materials Research Part A* 101 (2013): 1292-301.

Yin, Z., et al. “The Regulation of Tendon Stem Cell Differentiation by the Alignment of Nanofibers.” *Biomaterials* 31 (2010): 2163-75.

Yoshimoto, H., et al. “A Biodegradable Nanofiber Scaffold by Electrospinning and Its Potential for Bone Tissue Engineering.” *Biomaterials* 24 (2003): 2077-82.

Zhang, H. and Chen, Z. “Fabrication and Characterization of Electrospun PLGA/Mwnts/ Hydroxyapatite Biocomposite Scaffolds for Bone Tissue Engineering.” *Journal of Bioactive and Compatible Polymers* 25 (2010): 241-59.

Zhang, Y., et al. “Electrospun Biomimetic Nanocomposite Nanofibers of Hydroxyapatite/Chitosan for Bone Tissue Engineering.” *Biomaterials* 29 (2008): 4314-22.

Zheng, W., et al. “Endothelialization and Patency of RGD-Functionalized Vascular Grafts in a Rabbit Carotid Artery Model.” *Biomaterials* 33 (2012): 2880-91.

Zhu, X., et al. "Electrospun Fibrous Mats with High Porosity as Potential Scaffolds for Skin Tissue Engineering." *Biomacromolecules* 9 (2008): 1795-801.

Zilla, P., Bezuidenhout, D. and Human, P. "Prosthetic Vascular Grafts: Wrong Models, Wrong Questions and No Healing." *Biomaterials* 28 (2007): 5009-27.

Chapter 4

CONCLUSION

Electrospinning is a simple, proven technique to fabricate scaffolds from fibers on the nano- to micro-scale. Its many advantages include the ability to produce continuous nanoscale fibers, creating scaffolds with high surface area-to-volume ratios, and enabling the functionalization of fiber surfaces. In addition, the simplicity of the electrospinning process allows it to be easily scaled-up to industrial mass production. Furthermore, electrospinning allows many aspects of the resulting scaffold to be tailored to the specifications of the user. Fiber diameter can easily be controlled by changing the polymer concentration of the solution or altering the flow rate, the applied voltage, or the collecting distance. The nanoscale texture of the fibers can be manipulated to create different structures or patterns, such as nanopores and stripes. Nanofiber alignment can be controlled to create a rich variety of structures by increasing the rotational speed of the mandrel or by using specially designed collectors. Fibers can be spun from solutions containing multiple polymer types and additives, imparting them with a variety of tailorable properties. Many biomaterials can be electrospun, including natural polymers like collagen, elastin, chitin, chitosan, gelatin, and silk fibroin as well as synthetic polymers such as poly(lactic acid), polycaprolactone, poly(glycolic acid), polydioxanone, polyurethanes, and their copolymers. Small particles and bioactive materials can be encapsulated in the fibers by electrospinning, emulsion electrospinning, coaxial electrospinning, or post electrospinning modification. Because of their high porosity, large surface area, and appropriate fiber diameter and arrangement, electrospun scaffolds can mimic the structure of the native extracellular matrix. These properties allow for cell attachment and spreading as well as nutrient/waste transportation, making electrospun fibers well suited for tissue engineering applications.

Although remarkable progress has been made in the application of electrospun fibers in tissue engineering, there remain some challenges. One of the most significant drawbacks is the poor cell infiltration rate. While cell infiltration can be facilitated by the use of a sacrificial element to increase scaffold porosity, the mechanical integrity of the scaffold is often compromised. Furthermore, lack of stimulation could contribute to the cells remaining on the surface of the scaffolds rather than migrating deeper. Direct electrospinning of live cells has emerged as a novel technique to address this challenge, but there remain few published studies on this technique. Moreover, process variables controlling fiber morphology and functionality in applications of this new technique are yet to be studied, and there are no in-depth papers on its usefulness to clinical healthcare practice.

Great strides have been taken in the study of the electrospinning and its tissue engineering applications. However, the true importance of this technology will only be realized when electrospun nanofibers are commercialized and used clinically around the globe to improve human quality of life.

AUTHOR CONTACT INFORMATION

Xiumei Mo, PhD
Professor
State Key Lab for Modification of Chemical Fibers
and Polymer Materials, College of Chemistry
Chemical Engineering and Biotechnology
Donghua University, Shanghai, China
Email: xmm@dhu.edu.cn

Dawei Li
PhD Student
State Key Lab for Modification of Chemical Fibers
and Polymer Materials, College of Chemistry
Chemical Engineering and Biotechnology
Donghua University, Shanghai, China
and
Engineering Research Center of Technical Textiles
College of Textiles, Donghua University, Shanghai China

Benjamin Minden-Birkenmaier
PhD Student
Department of Biomedical Engineering
University of Memphis, Memphis, TN, US

Gary L. Bowlin, PhD
Professor and Herff Chair of Excellence
Department of Biomedical Engineering
University of Memphis, Memphis, TN, US

INDEX

F

H

I

L

M